Era		International Periods	Age in millions of years	Life developing overseas	Fossils in New Zealand	New Zealand Series
CENOZOIC	Quaternary	Recent	0.01	Man		Hawera
		Pleistocene	1.63	Ice Age		Wanganui
	Tertiary	Pliocene	5			Taranaki
		Miocene				Southland
			22.5			Pareora
		Oligocene	37.5	Modern mammals		Landon
		Eocene	53.5			Arnold
		Paleocene	65	Placental animals		Dannevirke
MESOZOIC		Cretaceous		Flowering Plants		Mata / Raukumara / Clarence / Taitai
			130			
		Jurassic		Reptiles and ammonites plentiful		Oteke / Kawhia / Herangi
			195			
		Triassic		Early ammonites and dinosaurs, primitive mammals		Balfour / Gore
			235			
PALEOZOIC		Permian	280	Reptiles develop		
		Carboniferous	355	Coal forests, insects		
		Devonian	413	Fishes common		
		Silurian	425	First land plants		
		Ordovician	475	Graptolites		
		Cambrian	600	Animals with hard shells		
		Precambrian	4600?	Solidification of earth		

Field Guide to
New Zealand
GEOLOGY

Also by Jocelyn Thornton:
  Mobil New Zealand Nature Series: Gemstones

Field Guide to New Zealand GEOLOGY

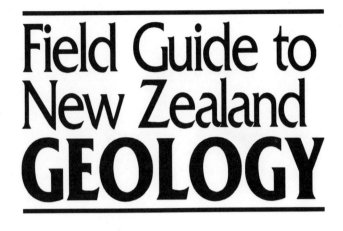

An Introduction to Rocks, Minerals and Fossils

Jocelyn Thornton

Heinemann Reed

Published by Heinemann Reed
a division of Octopus Publishing Group (NZ) Ltd
39 Rawene Road, Birkenhead, Auckland. Associated
companies, branches and representatives throughout
the world.

ISBN 0 7900 0025 3

First published 1985
Reprinted 1988

Cover design: Cathy Clear
Cover photography: Jocelyn Thornton
Printed by Kyodo-Shing Loong Printing Ind. Pte. Ltd.

Contents

Acknowledgements

I owe many thanks to the geologists who have published their findings and others who patiently answered questions. I am especially grateful to all those who organised field trips and wrote guidebooks. Officers of the Geological Survey have been most helpful, and many of the maps and diagrams have been based on their publications. Full acknowledgements are given in the notes on the illustrations at the end of the book. The Geology Department at Victoria University helped by allowing me the use of drawing and photographic facilities. I thank Jim Cole, John Collen, Roger Cooper and Paul Vella, who read the text and helped with suggestions, although I take full responsibility for my interpretations of the rocks. Janene McDermott at Reed Methuen edited the text and Lindsay Cuthbertson put text and pictures in harmony. I am grateful for their help.

Finally, I thank my family for sharing in my rock hunting and for their acceptance of a house full of rocks.

Introduction

Anyone who has travelled in New Zealand will have seen how the variety of our landscape depends on the rocks lying beneath the plant cover, from the volcanic cones of Auckland to the schist tors of Central Otago. Many of us pick up pebbles along the way and wonder what they are and how they were made. This book attempts to provide answers in the simplest of terms and with illustrations of New Zealand's more common rocks, minerals and fossils and the places where they can be found.

The first chapter explains very briefly how rocks are formed and outlines the structure of the earth and our understanding of its geological history. The rest of the book describes the rocks of New Zealand, using the geological method of starting at the bottom of the pile with the oldest rocks.

In order to show the growth and development of our land, general description has been supplemented with specific examples. Geology is a very practical science, and rocks, unlike living creatures, stay still while we look at them — although the speed with which some roadcuts are concealed by vegetation can be very distressing. This book contains many detailed descriptions of areas where the rocks themselves illustrate their story, to enable the reader to imagine areas as they were when the rocks were forming — perhaps a quiet sea floor, or a lush swamp or the scene of a volcanic eruption. Our rocks were created in many ways and our land has changed many times in the 600 million years that it has been in the making.

Beginners sometimes fear geological jargon, but most of the processes involved in geology may be explained in simple language (any geologists reading this book will have to supply their own translations). The names of rocks and minerals in the text are introduced gradually, and the names of the geological time periods are explained with reference to their origins. A copy of the time scale is reproduced inside the cover as a ready reminder.

The diagrams in this book should be read as part of the text, especially in the first chapter. The landscapes were mostly photographed from the road or beach and show formations accessible to the average traveller. In a number of the maps the orientation is south at the head of the page to match the viewpoint of a drawing or photograph that accompanies the map.

The first two maps are intended as a quick reference for travellers. They show the main roads and the page numbers where descriptions of the types of rocks in those areas appear. Although the book was begun with the intention of discussing mostly the areas where the rocks from each period of our history were first recognised, it soon became clear that the rocks of many other parts of the country could illustrate incidents in our history and that travellers could make good use of a guide to the rocks in other regions.

It has taken me a long time and many examples to develop even a poor eye for the main rock sequences. I hope the repetition of examples in this book will help other beginners to recognise the rocks, wherever they may be, and the extra details provided may suggest places of interest for detours.

Handle our rocks with care and let them tell their own story.

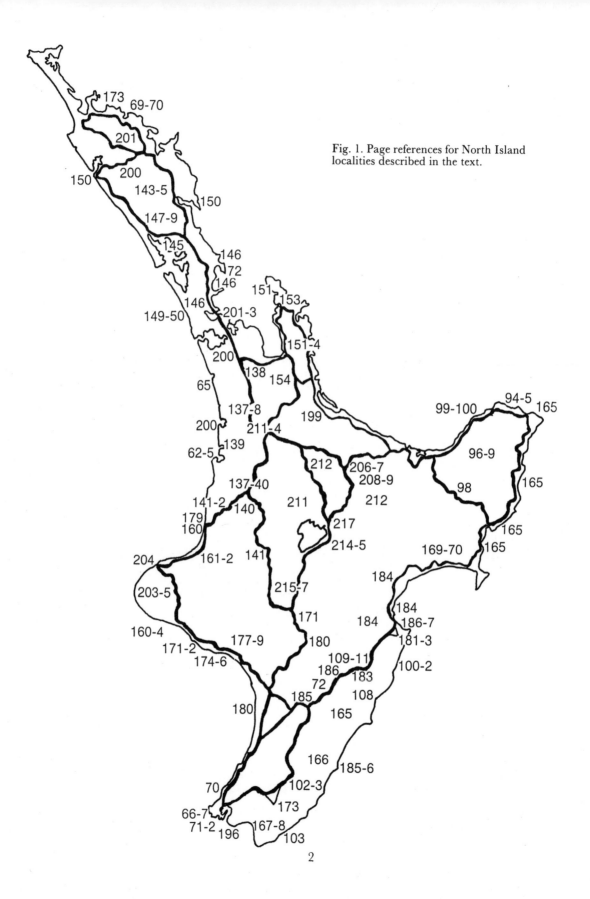

Fig. 1. Page references for North Island localities described in the text.

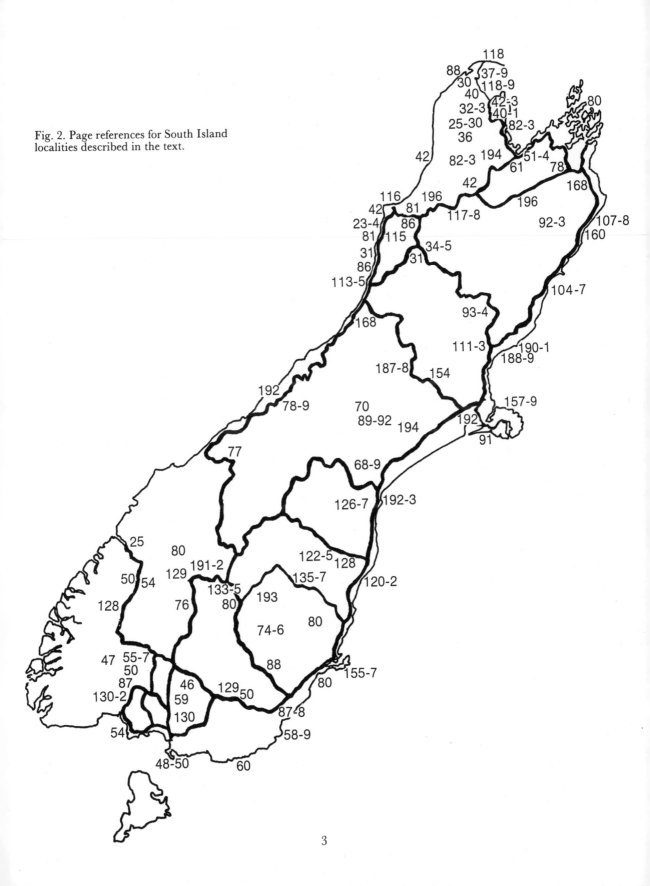

Fig. 2. Page references for South Island localities described in the text.

118
88 37-9
30 118-9
40 42-3
32-3 40-1
25-30 82-3
36
42 82-3 194 51-4
61 78
42 168
116 196
42 81 117-8 196
23-4 86 92-3 107-8
81 115 160
31 34-5
86 31
113-5 104-7
168 93-4
111-3 190-1
188-9
187-8 154
192 157-9
78-9 70 192
89-92 194 91
77 68-9
126-7 192-3
25
80 122-5 128
50 54 129 191-2 135-7 120-2
133-5 193
128 76 80 80
74-6
47 55-7 88
50 80 155-7
87 46 129 50
130-2 59
130 87-8
54 58-9
48-50 60

CHAPTER ONE

The groundworks:
An introduction to geology

LOOKING at our country we can see that the landscape is not permanent, that the rocks are moving all the time as every passing storm sends mud and stone a little further on their journey from the mountains to the sea.

Since the formation of the earth some 4,600 million years ago, the rocks have been continuously reworked. Once they are exposed to the weather they break down. The pieces may be cemented together, changed as new crystals grow, or melted completely to start again as new rocks. This is known as the rock cycle (Fig. 3).

If erosion and gravity carry rocks from the heights to the depths of the earth it must take strong forces within the earth to lift them back up again. These forces are formed by heat

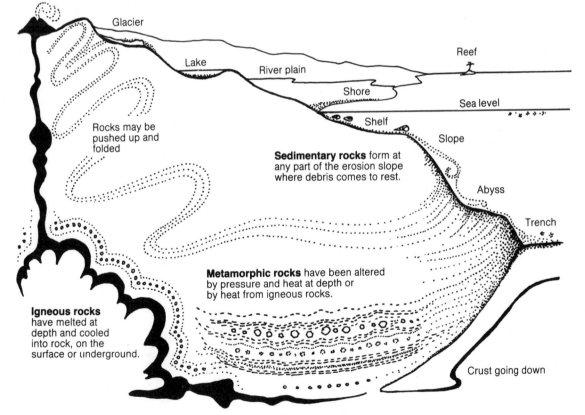

Fig. 3. The rock cycle.

flowing from the centre of the earth and concentrating into areas of greater heat and energy as it makes its way up through the various layers (Fig. 4).

The topmost layer, the crust, is like a jigsaw puzzle, divided into interlocking plates which rest on a slippery layer which is probably partially melted. The plates jostle and move like scum on simmering jam.

The joints between the plates are mainly of two kinds — spreading centres and collision zones (Fig. 5). At spreading centres rising hot

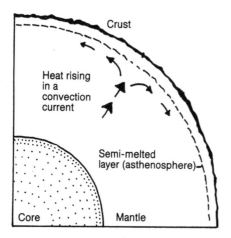

Fig. 4. Inside the earth.

currents bring melted material to the surface to form a belt of new crust as the older crust moves apart, sliding over the slippery layer. The far edges of these plates butt against each other, and at collison zones old crust at the edges of the growing plates is destroyed by being pushed down beneath a neighbouring plate. In some places the edges of the plates simply slide sideways past each other. In this way the materials of the crustal area are constantly recycled (Fig. 6).

Oceanic crust is relatively thin, measuring about 5-8 kilometres in depth. It is made of a dark, heavy rock rich in iron, magnesium and silica. It is sometimes called sima (from the first two letters of two of its components, *si*lica and *ma*gnesium).

Crust under continents is up to 35 kilometres thick and is mostly composed of rock rich in *si*lica and *al*uminium. It is known as sial. This rock is lighter than sima. When two neighbouring plates collide sial may therefore rumple up, but it cannot be dragged down and destroyed because it floats on sima like oil on water. Only oceanic crust can sink to where it is gradually absorbed back into the subcrustal material from which it originated (the mantle); or it can be melted in the friction of a collision. Continental crust can also melt in the friction of a collision, usually at the edges of continents just above collision zones (Fig. 7).

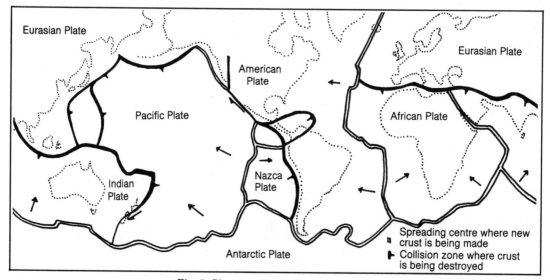

Fig. 5. Plates on the surface of the earth.

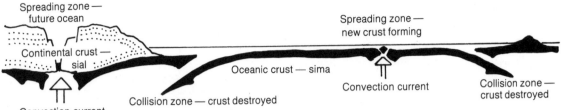

Fig. 6. Section across the crustal plates. The arrow-heads drawn at plate boundaries indicate movement and direction, not the shape of the plate edge.

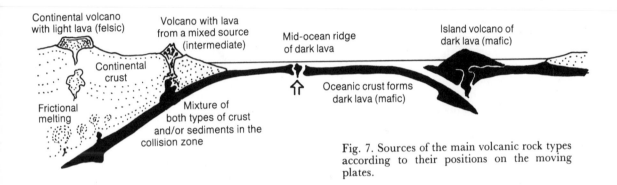

Fig. 7. Sources of the main volcanic rock types according to their positions on the moving plates.

Igneous rocks are formed by the melting of the earth's crust. They vary in composition according to their origins. Oceanic crust makes a magma (molten material) rich in iron and magnesium; it is dark and heavy when cooled into rock. Continental crust melts to make a pale, lighter magma rich in silica and aluminium. It is described as felsic (it contains *fel*dspars and *si*lica). Mixtures of the two crusts at a collison zone make an intermediate magma.

After melting, the materials rise towards the surface and begin to cool. The elements assemble into groups, and as long as the magma remains a liquid the atoms can move to join the growing clusters. Solid clusters of atoms with a regular, distinctive structure form the crystals of minerals, and each mineral has its own set of elements.

Although there are about 3,000 known minerals, nearly all our rocks are made from a small number of what are called the rock-forming minerals. These are mostly silicates, containing clusters of one silicon atom surrounded by four oxygen atoms (making a closely packed pyramid), linked with various other elements (Fig. 8).

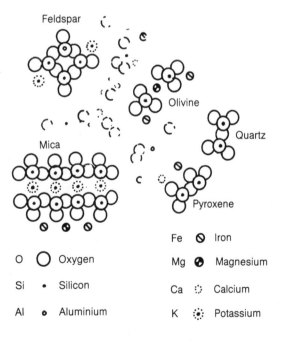

O	◯ Oxygen		Fe	◉ Iron
Si	• Silicon		Mg	◉ Magnesium
Al	∘ Aluminium		Ca	⋮ Calcium
			K	⋮ Potassium

Fig. 8. Atoms in a liquid linking into groups to make minerals (in other words, silicates crystallising in a magma).

In any magma some minerals start crystallising early at very high temperatures — up to 1300°C — growing small, well-shaped crystals which float in the molten rock. Usually these minerals use up or gather into their structure any magnesium, iron or calcium that is in the magma mix, whether it be a little or a lot, so minerals formed later contain lesser amounts of these elements.

If the magma comes right to the surface of the earth and erupts from a volcano it is called lava. Because lava cools quickly it sets hard into rock and the minerals have to stop growing immediately; they are frozen solid. Therefore volcanic rocks may contain only a few visible crystals of the early-forming minerals set at random in a very fine-grained rock.

But if the magma remains underground it cools very slowly and the crystals have time to grow until all the magma is used up (Fig. 9).

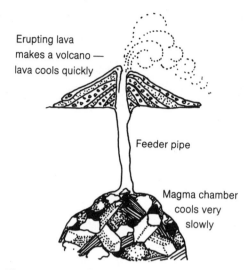

Erupting lava makes a volcano — lava cools quickly

Feeder pipe

Magma chamber cools very slowly

Fig. 9. Formation and texture of volcanic and plutonic rocks.

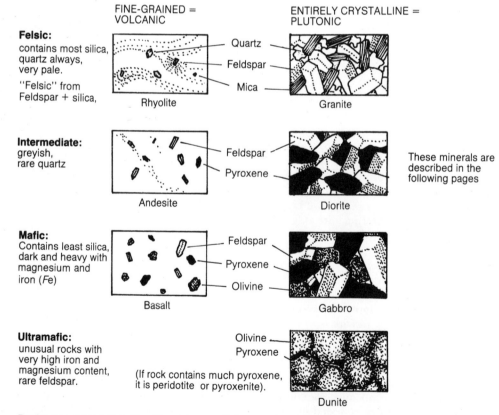

FINE-GRAINED = VOLCANIC

ENTIRELY CRYSTALLINE = PLUTONIC

Felsic:
contains most silica, quartz always, very pale.

"Felsic" from Feldspar + silica,

Quartz
Feldspar
Mica

Rhyolite

Granite

Intermediate:
greyish, rare quartz

Feldspar
Pyroxene

Andesite

Diorite

These minerals are described in the following pages

Mafic:
Contains least silica, dark and heavy with magnesium and iron (Fe)

Feldspar
Pyroxene
Olivine

Basalt

Gabbro

Ultramafic:
unusual rocks with very high iron and magnesium content, rare feldspar.

Olivine
Pyroxene

(If rock contains much pyroxene, it is peridotite or pyroxenite).

Dunite

Dunite was named from Dun Mountain near Nelson, and peridotite is from the gem variety of olivine, peridot, which is the main mineral present in these rocks, although not in gem quality.

Fig. 10. Main classes of igneous rock.

They may become as large as fingernails. Thus the larger the crystals in igneous rock the slower the rock has cooled. With this knowledge we can readily identify igneous rock as volcanic (formed from erupted lava) or plutonic (formed from magma, underground). The latter name is derived from Pluto, the Roman god of the underworld.

All igneous rocks are categorised according to their composition and their grain size, which is determined by their pace of cooling and shows whether they are volcanic or plutonic. Fig. 10 illustrates the commonest categories, but there are many varieties according to the proportions of the different minerals that form these rocks.

THE ROCK-FORMING MINERALS

When geologists look at minerals they use small magnifying glasses or pocket lenses, and pocket knives are used to test for hardness. The degree of hardness and the property of cleavage are used to identify various minerals. (Cleavage is the way in which some minerals split along planes of weak atomic bonding. [Fig. 11.] It produces clean, flat surfaces on a broken crystal or a series of tiny, parallel surfaces which simultaneously reflect light when the crystal is turned.)

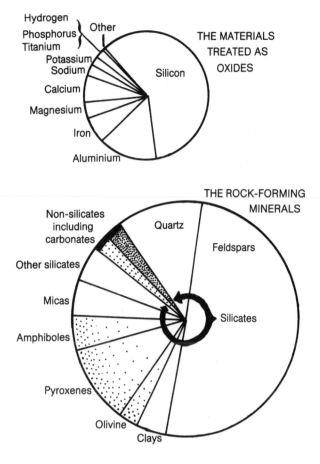

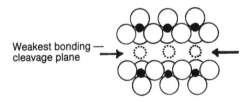

Fig. 11. Cleavage planes.

When minerals have grown in solid rock their edges interlock and they cannot form the clean outer walls called crystal faces. A perfectly shaped crystal has to grow with liquid or a space around it; this occurs most often in rocks which have cavities. Illustrations usually show perfect crystals that have grown under ideal conditions, but these are not always easy to find. Crystal faces should never be used to test for hardness because their natural surfaces must not be spoilt. A broken surface should be used both to test for hardness and to view cleavage.

Fig. 12. Proportion of materials and minerals in the earth's crust.

Because of the proportions of the elements in the earth's crust, some 90 per cent of our rocks are formed from silicate minerals. These are discussed in the following pages. Nine times out of ten the minerals you find will be of this type (Fig. 12). Most silicates are harder than steel and cannot be scratched with a knife blade.

Quartz

Quartz is pure silica, one of the commonest minerals. Its finest crystals are like glass, which is not surprising because glass is made by melting and cooling silica sand; but quartz can also be opaque, white or coloured by impurities. Purple quartz is called amethyst; it is produced by radiation on traces of iron. Where quartz crystals occur they are six-sided, and their pointed ends (terminations) are generally made up of three larger sides and three smaller triangles. Quartz cannot be scratched with a knife. It is the one common, pale mineral that has no cleavage but instead a rough, broken surface or curved (conchoidal) fracture surfaces (Fig. 13).

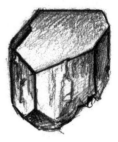

Fig. 15. Orthoclase feldspar, from Paihia, Southland.

Fig. 13. Quartz crystals.

Feldspars

Feldspars, by contrast, have excellent cleavage and appear as flat, blocky masses in coarse-grained igneous rocks. They are even more abundant than quartz in the rocks of the world, but most feldspars are too small to see in the groundmass of our volcanic rocks.

Feldspars are a group of minerals with similar structure but variable composition, which determines their individual names. They all contain aluminium and silica and another one or two elements out of potassium, sodium and calcium. Fig. 14 shows the names of the feldspars arranged according to their composition (those containing the most sodium are at the left while those containing the most calcium are on the right).

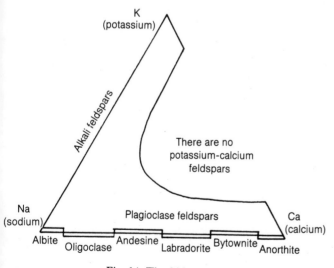

There are no potassium-calcium feldspars

Fig. 14. The feldspar group.

Orthoclase, the most common K(potassium)-feldspar, is shaped like a brick pushed backwards, leaving its front and top crystal faces squarish. The name comes from the Greek "ortho", meaning square, upright (Fig. 15). Plagioclase is a group name for all the sodium-calcium feldspars, and their crystal shape is that of a brick squashed completely out of square so all faces are skewed (from the Greek "plagios", meaning slanted).

Plagioclase feldspars grade from albite, which contains 90-100 per cent sodium, to anorthite, which is 90-100 per cent calcium. The percentages and resulting names are determined by chemical analysis or by using a thin slice of rock and an especially designed microscope for identifying minerals from their optical properties.

Feldspars containing the alkali metals potassium and sodium are called alkali feldspars; they have no series of mixed composition like plagioclases, but there can be a mix of interlayered orthoclase and albite, which is known as perthite.

Generally we simply refer to feldspars, or orthoclase and plagioclase feldspars, or K-feldspars and plagioclase.

Orthoclase, especially the salmon-pink variety, is common in granites and fairly obvious; but white or creamy feldspars are likely to be mistaken for quartz until you look closely for the cleavage planes on the broken surfaces. You may need a hand lens for this, especially if viewing the small crystals in volcanic rocks.

Sometimes crystals of feldspar grow in fine layers of two different compositions. In this case light will reflect off the layers, giving a sheen or "schiller" effect. Moonstones are feldspars, as are the shining bluish crystals in some dark, ornamental building stones.

Micas

Micas (Fig. 16) are flaky minerals which can be split into paper-thin layers because of their perfect cleavage in one direction only. They are also fairly soft; flakes of mica can be shattered by a knife point. Because of its cleavage mica does not conduct electricity and has been used extensively inside toaster elements. Black mica (biotite) contains iron and magnesium. Flakes of the less dark varieties can be mistaken for gold, but they can be easily distinguished because gold will resist a knife point. White mica was once used for window panes in Moscow, so it is now called muscovite.

Fig. 16. Mica.

Amphiboles and pyroxenes

These are two closely related groups of iron-magnesium silicates with black or greenish crystals. The atomic structures within each group are similar, but their compositions vary and some of the minerals are very hard to distinguish unless viewed under a microscope. Both groups are opaque, non-metallic, harder than steel and show many fine cleavage surfaces.

Pyroxenes (Fig. 17) tend to have blocky crystals and are square in cross-section. Where the cleavage planes are visible, they intersect at right-angles. Pyroxenes include diopside and enstatite (green and brown minerals rich in magnesium) and the iron-rich augite, which can range in colour from green to black.

Fig. 17. Augite from Oakura Beach, Taranaki, showing the crystal and a top view of the typical blocky pyroxene shape and 90-degree cleavage planes.

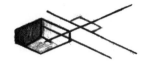

Fig. 18. Hornblende from Oakura Beach, showing the crystal and top view with the amphibole cleavage planes at about 60 and 120 degrees.

Amphiboles (Fig. 18) tend to have long-bladed crystals with a cleavage angle of 124 degrees. Hornblende is the most common amphibole in both igneous and metamorphic rocks. Other amphiboles include the golden-brown cummingtonite, white tremolite and green actinolite, all likely to be found in the West Coast rivers. A dark-blue variety, riebeck-ite, occurs in Douglas Creek on the Haast River in very small crystals. Both hornblende and augite crystals can be found weathering out of the cliffs at Oakura, south of New Plymouth. It is an interesting exercise to collect both.

Garnets

Garnets also form a group of minerals with a common structure but variable chemistry. They have no cleavage and are commonly red with a good crystal shape (Fig. 19); but we have here also a massive green garnet — grossular. It is popular as a gemstone (Plate 1E).

Fig. 19. Crystal shapes of garnets.

Olivine

Olivine, the gem peridot, is an olive-green mineral found as crystals in mafic rocks or making up dunite (page 53, also Plate 1B). The lime-green mineral found in some metamorphic rocks and beach pebbles is more likely to be epidote, from another silicate group.

11

Calcite

This is the main non-silicate rock-forming mineral. It is found occasionally in igneous rocks but is an essential component of the limestone group of sediments (Plate 5B). Calcite crystals are relatively soft. They are easily scratched by steel or a copper coin (which distinguishes them from feldspars) and they have excellent cleavage in three directions which makes them form blocky rhombs when broken (Fig. 20). The chemical name for calcite is

Fig. 20. Calcite crystals and broken block showing rhombic cleavage planes.

calcium carbonate — a calcium atom is linked to the carbonate group of one carbon and three oxygen atoms. When acid is added to a carbonate, carbon dioxide is released and the acid-mineral fizzes. This is a test for calcite, any limestone and many other carbonates such as the blue and green copper carbonates.

The minerals most likely to be seen therefore are pale feldspars, quartz and calcite, platy micas and the green to black iron magnesium minerals such as hornblende, augite and olivine. These make up most of our rocks.

GEOLOGICAL TIME

If rocks are continually being recycled, with periods of uplift followed by periods when sediments are being formed, we need to have some idea of the time at which particular events took place and various groups of rocks were being made. Human time is measured by a sequence of physical events — the swinging of a pendulum, the rotation of the earth and the phases of the moon. We can see time recorded in the growth rings of trees and shells and in our memories ("When you were little . . ."). However, geological time is immense. People first began to appreciate its vastness when they saw that some of the rocks of the earth were made up of layers of sediments, that each layer represented centuries of slow build-up of sands and muds and that some of the layers had been tilted and worn away before other layers were piled on top (Fig. 21).

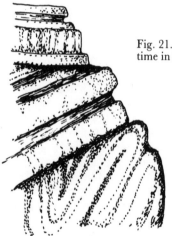

Fig. 21. The record of time in layers of rock.

Our first modern histories of the earth and the first time charts were written in Great Britain where geologists looked at the tilted layers of rocks found across the country from Wales to London (Fig. 22). On paper they untilted the layers, recognising that those at the bottom had to be the oldest, and drew up a diagram of the sequence of layers or strata known as a stratigraphic column (Fig. 23). These columns always have the oldest rocks at the bottom and they also use conventional symbols for various types of rock.

The geologists then began to work out the time scale represented by this column. They

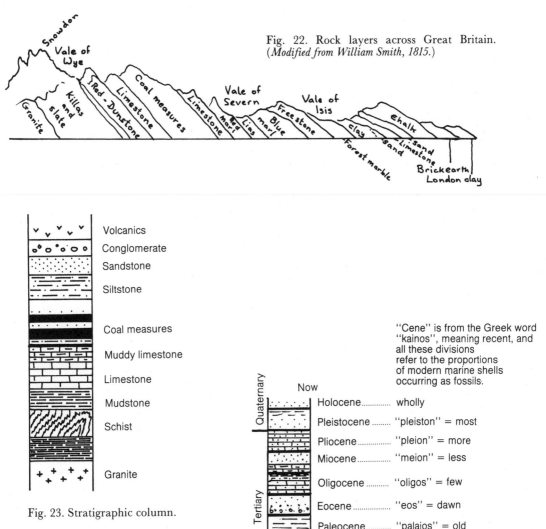

Fig. 22. Rock layers across Great Britain. (*Modified from William Smith, 1815.*)

Fig. 23. Stratigraphic column.

"Cene" is from the Greek word "kainos", meaning recent, and all these divisions refer to the proportions of modern marine shells occurring as fossils.

Holocene	wholly
Pleistocene	"pleiston" = most
Pliocene	"pleion" = more
Miocene	"meion" = less
Oligocene	"oligos" = few
Eocene	"eos" = dawn
Paleocene	"palaios" = old
Cretaceous	"creta" = chalk (the white cliffs of Dover)
Jurassic	Jura Mountains
Triassic	threefold division (in Germany)
Permian	Perm, a town on the Volga River
Carboniferous	coal-bearing
Devonian	Devon
Silurian	Silures, Celtic tribe of the borders
Ordovician	Ordovices, Celtic tribe of Wales
Cambrian	Cambria, Roman name for Wales

Fig. 24. The names of the periods and epochs and their derivations.

named the periods of time after the areas where the rocks formed during each period were first recognised and described. The oldest sedimentary rocks containing fossils (see page 14) were called Cambrian, for Wales. Cambrian rocks are now known to be some 500 to 650 million years old.

In their earliest attempts to classify rocks geologists used the labels Primary, Secondary, Tertiary and Quaternary, with Primary applying to the crystalline oldest rocks, Secondary to older sediments and so on. Although the terms Primary and Secondary are no longer used, Tertiary and Quaternary are still used to describe the rocks of the last 65 million years (Fig. 24).

Rocks older than Cambrian were simply called Precambrian, and at first the period of time represented by the Precambrian rocks was thought to be relatively short. Gradually geologists have discovered the extremely long period of time since the earth first solidified, and now Precambrian time is considered to be 4,000 million years.

The first stratigraphic columns were made up based solely on the various rock types. Many of them were very distinctive; for example, layers of black slates, red sandstones, white chalk or seams of coal. All of these rocks are sediments and many contained fossils.

Fossils also record time

A fossil is any remains of a living organism preserved in rock; it can be footprints or just dung. It is very rare for the whole creature to be preserved because usually only the hard parts such as shell or bone survive burial. Sometimes these, too, have disappeared leaving only their imprint. The Cambrian rocks contained many fossils of very primitive shelled animals, but younger rocks were found to contain fossils of more complicated creatures, giant trees and animals with bony skeletons, fishes and reptiles.

Initially, church people taught that each group of animals was specially created and that any group not present today had died out in the flood or in other catastrophies. But gradually the theory of evolution of life prevailed, after Charles Darwin explained a method of change and adaptation that allowed creatures to pass on to their offspring the special characteristics that enabled them to survive when their less fit relatives died.

Life on earth seems to have evolved from simple cells living in the sea to more complicated organisms which gradually moved on to the land and later into the air (Fig. 25). In the long procession some groups died out completely and others developed into new forms. For the geologist the changing pattern of evolution is a record of the passing of time.

As each layer of rock is laid down it may contain the remains of some distinctive creature that will identify that timespan wherever that creature is found, in whatever kind of rock. A similar fossil found in sandstones formed in one country and mudstones found in another may link the two types of rock in time. Such fossils are called index fossils. The most valuable index

Fig. 25. The procession of life on earth.

fossils are those that spread over a wide area, perhaps across the oceans, yet lived for a relatively short period of time before developing some new characteristic, becoming a new species and thus defining a new period.

As soon as the value of index fossils was realised, paleontologists (people who study old life forms) or fossil collectors moved across the continents hunting fossils, using them to link and relate various outcrops of rock. Gradually they discovered a sequence of life forms similar to the column of rock types (Fig. 26). Now the whole of geologic time is divided according to the dominant life forms of the periods. The Phanerozoic (meaning obvious life), which extends from the Cambrian period to the present, is further divided into the three eras — Paleozoic (ancient life), Mesozoic (middle life) and Cenozoic (recent life).

Fossils can thus be used to fit rocks into the geological time sequence.

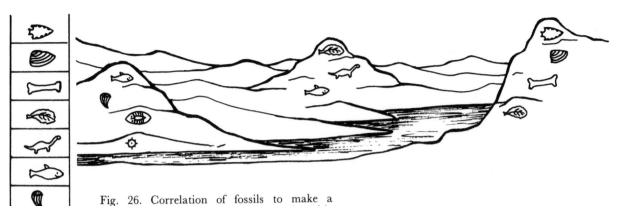

Fig. 26. Correlation of fossils to make a sequence of life forms — a bio-stratigraphic column.

Radiometric clocks

Finally nuclear science gave us a more precise dating method, based on the fact that some elements are unstable and "decay", changing by radioactivity into new, stable elements. From the time an igneous rock cools and the elements crystallise into minerals, the unstable minerals start to alter. The changing proportions of the old and new elements tell us the time that has passed since the start of the decaying process.

Elements used in radiometric dating include potassium[40], rubidium[87], uranium[238] and carbon[14]. We will use potassium as an example. This element is distinguished by its 19 protons, added to which is a varying number of neutrons; the total of the two is used to identify the potassium isotopes. Potassium[39] is stable, but potassium[40] has an extra neutron and is unstable or radioactive, and it decays to form argon[40].

Potassium occurs in micas and some feldspars, so these minerals are extracted from the rock for dating purposes, or sometimes samples of the entire rock containing these minerals are used. Then the relative amounts of potassium[40] and argon[40] are determined. The rate of alteration or decay is measured in terms of "half-lives" — the time half the original or parent element takes to alter to the new or daughter element. In each successive half-life, half the remaining radioactive element will alter, so the actual rate of decay tapers off with time. The graph of the decaying process is the same for all elements, as shown in Fig. 27. Because we know the half-lives of the unstable elements (potassium[40] has a half-life of 1,300 million years, and carbon[14] 5,730 years) we can calculate the age of the rock. The number of half-lives that have passed (based on the proportion of the elements present) is multiplied by the extent of the known half-life, giving the age of the rock in years.

By dating igneous rocks at appropriate places in the time-rock column geologists can fix the ages of the geological time periods. The first attempts at radiometric dating were very approximate, but modern techniques are continually providing more accurate results and the ages have been adjusted accordingly. They are now presented as the International Geological Time Scale which includes boundary dates and index fossils.

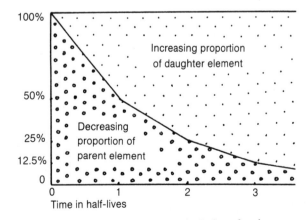

Fig. 27. Graph for radiometric dating of rocks.

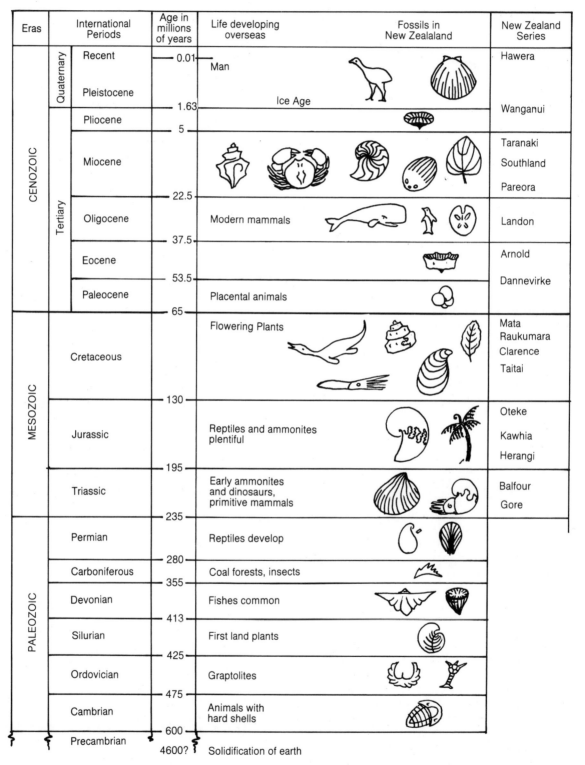

Eras	International Periods		Age in millions of years	Life developing overseas	Fossils in New Zealaland	New Zealand Series
CENOZOIC	Quaternary	Recent	— 0.01 —	Man		Hawera
		Pleistocene	— 1.63	Ice Age		Wanganui
	Tertiary	Pliocene	5 —			
		Miocene	— 22.5			Taranaki Southland Pareora
		Oligocene	— 37.5	Modern mammals		Landon
		Eocene	— 53.5			Arnold
		Paleocene	— 65	Placental animals		Dannevirke
MESOZOIC	Cretaceous		— 130	Flowering Plants		Mata Raukumara Clarence Taitai
	Jurassic		— 195	Reptiles and ammonites plentiful		Oteke Kawhia Herangi
	Triassic		— 235	Early ammonites and dinosaurs, primitive mammals		Balfour Gore
PALEOZOIC	Permian		— 280	Reptiles develop		
	Carboniferous		— 355	Coal forests, insects		
	Devonian		— 413	Fishes common		
	Silurian		— 425	First land plants		
	Ordovician		— 475	Graptolites		
	Cambrian		— 600	Animals with hard shells		
	Precambrian		4600?	Solidification of earth		

Fig. 28. The geological time scale.

But here in New Zealand we have local names to add as well (Fig. 28).

The New Zealand Series

Because New Zealand is so far from Europe, many of its rocks and fossils are very different. The sequence of rocks and fossils must be recognised and defined by fossils peculiar to New Zealand (these have been correlated occasionally by the occurrence of species also found overseas) and local names must be added to the scale to make it relevant to these findings. These names are known as the New Zealand Series. They arose as follows.

As geologists found good examples of New Zealand rocks representing each geological age they built up a sequence of "type sections" with fossils and sediments which became a standard reference for comparison with all other rocks of that age in the country. For example, rocks in the banks of the Waiau River in Southland are the type section for all New Zealand rocks of the middle Miocene age; "Southland" is therefore listed in the New Zealand Series of the geological time scale opposite the middle Miocene period. We talk about the Southland Series, meaning the rocks typical of that area, and we refer to Southland time, meaning the time when these rocks were being formed. The Southland Epoch of time is subdivided into ages whose rocks are known as the Southland Stages — the Altonian Stage, the Clifdenian Stage, the Lillburnian Stage and the Waiauan Stage; all are named after localities in the area.

These and other New Zealand Series names can easily be forgotten or ignored, but they are used on the 1:250,000 or 4-mile-to-the-inch geological maps that are indispensable for anyone seriously looking at the geology of New Zealand (see page 218 for details of obtaining them). Every part of these maps — whether North Auckland or Stewart Island — has its divisions marked with series and stage initials (for example, Sc, Sl and Sw) and although they

are explained in the key at the side of each map it is handy to know and remember them. Since the names will be meaningful only after you have seen the places they refer to, pictures and maps of the type sections are given in this book. Visitors to type section areas must remember that these areas are valuable reference material and must not be damaged or despoiled.

THE EVOLUTION OF NEW ZEALAND

The New Zealand time scale arises from our isolation from Europe and North America, as already stated, and also from our different geological history, caused by our position on the jigsaw of crustal plates.

The large continents have central areas that are very old and rigid and are known as shields. Erosion of these shields gave rise to seaward borders of sediments which were then pushed up and stuck on to the shields, and the continents grew as the process was repeated (Fig. 29).

However, New Zealand has no such shield area. Our land developed as part of the border of a continental shield composed of Antarctica and Australia when those two continents were part of the great southern super-continent, Gondwana (Fig. 30). Gondwana later broke up when the crustal plates moved apart.

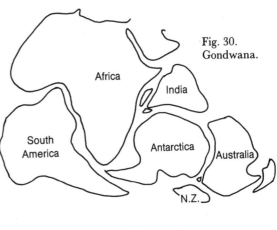

Fig. 30. Gondwana.

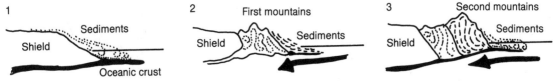

Fig. 29. The growth of continents.

17

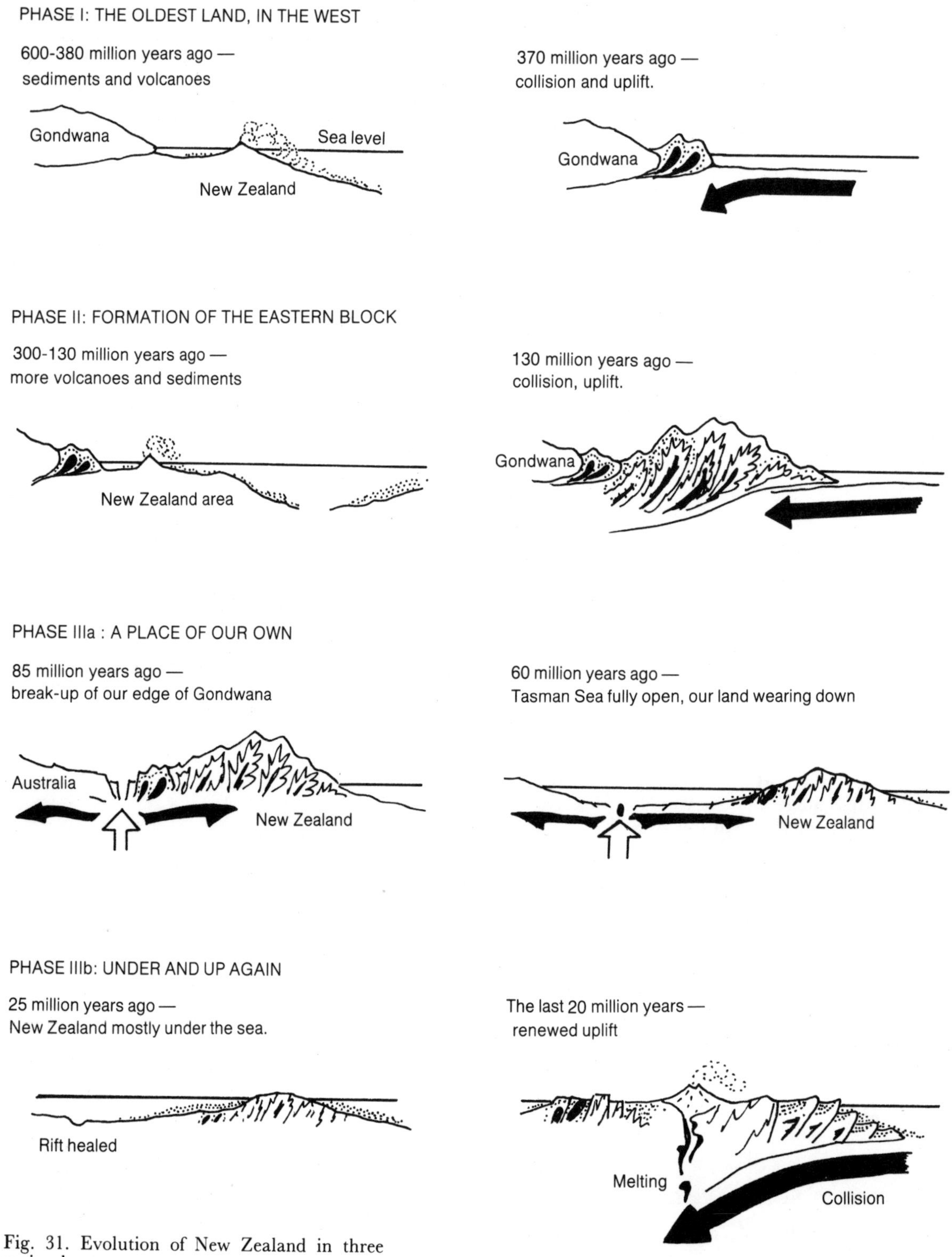

PHASE I: THE OLDEST LAND, IN THE WEST

600-380 million years ago —
sediments and volcanoes

370 million years ago —
collision and uplift.

Gondwana

Sea level

New Zealand

Gondwana

PHASE II: FORMATION OF THE EASTERN BLOCK

300-130 million years ago —
more volcanoes and sediments

130 million years ago —
collision, uplift.

New Zealand area

Gondwana

PHASE IIIa : A PLACE OF OUR OWN

85 million years ago —
break-up of our edge of Gondwana

60 million years ago —
Tasman Sea fully open, our land wearing down

Australia

New Zealand

New Zealand

PHASE IIIb: UNDER AND UP AGAIN

25 million years ago —
New Zealand mostly under the sea.

The last 20 million years —
renewed uplift

Rift healed

Melting

Collision

Fig. 31. Evolution of New Zealand in three main phases.

18

For most of our history New Zealand was at the edge of the ocean, continually pressured by the moving crust beneath. Like most of the border areas, New Zealand is considerably younger than the adjacent shield areas (in this country there have been discovered to date only a few small areas that seem to go back to the Precambrian Period). Our whole geological history can be divided into three main periods of sedimentation and uplift as shown in Fig. 31, punctuated by our departure from the sheltering Gondwana shield.

By the end of the first phase in our history the rocks of the Western Province were formed. This is sometimes called the Tuhua sequence, from Mount Tuhua near Lake Kaniere — a mountain of granite formed in the uplift at the end of the first phase. At the end of the second phase the rest of the foundations of New Zealand were in place (Fig. 32). The rocks formed at this stage

are sometimes called the Rangitata sequence, from the Rangitata River where the rocks of this age group are well displayed.

Then began the third phase, which still continues, in which a blanket of softer sediments was draped over the older rocks and we split from Australia and became established as a small continental area through which a plate boundary passes (Fig. 33). The last upheaval shaped the Kaikouras, therefore the third phase is often referred to as the Kaikoura sequence.

Our Alpine Fault, marked on the map (Fig. 33) and clearly visible in the Landsat photo-

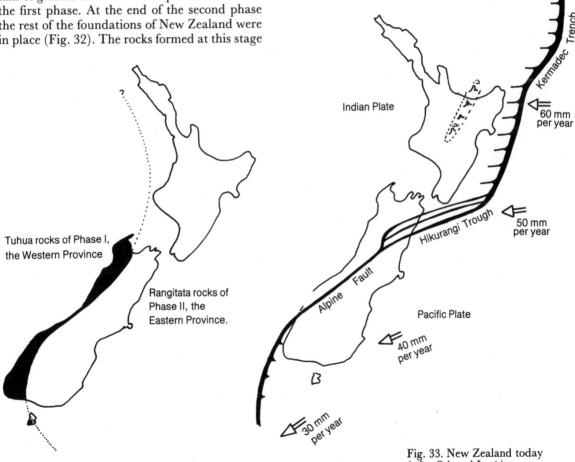

Fig. 32. Basement rocks in New Zealand.

Tuhua rocks of Phase I, the Western Province

Rangitata rocks of Phase II, the Eastern Province.

Indian Plate

Kermadec Trench

60 mm per year

50 mm per year

Hikurangi Trough

Alpine Fault

Pacific Plate

40 mm per year

30 mm per year

Fig. 33. New Zealand today (*after Cole and Lewis*).

Fig. 34. The Alpine Fault. (*Original Landsat imagery obtained from NASA and prepared by the Lands and Survey Department in co-operation with Physics & Engineering Laboratory, DSIR.*)

Fig. 35. Small fault cutting sandstone, Island Bay, Wellington.

graph (Fig. 34) is a major break in the crust where the Indian and Pacific Plates slide past each other. A fault is any break where differential movement is or once was taking place. It may be sideways movement, vertical movement or, as in New Zealand, any combination of the two. It can sometimes be seen on a very small scale, as in the break in the rock layers shown in Fig. 35.

This chapter was intended to provide a very general introduction to geological processes and the formation of New Zealand. The next chapters deal in greater detail with the rocks of each region of the country and the events they record. There is one chapter for each major phase of our geological history, beginning with the oldest.

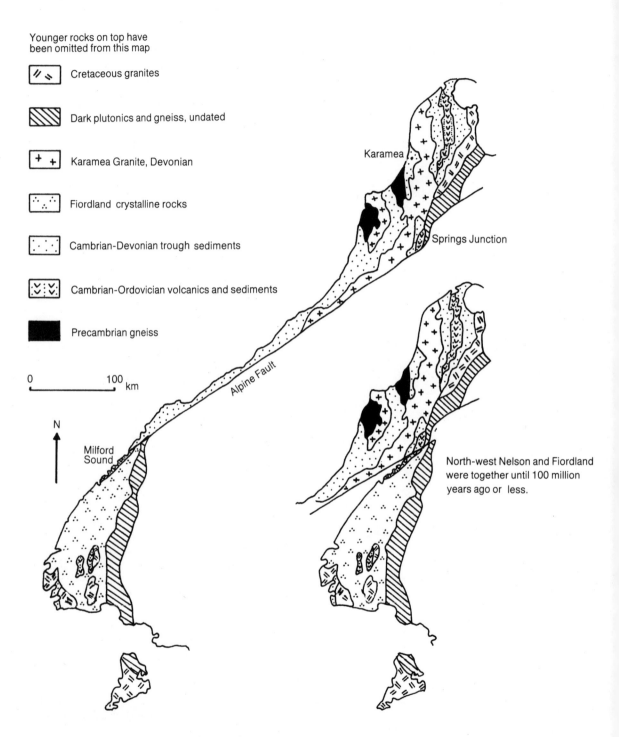

Younger rocks on top have been omitted from this map

Cretaceous granites

Dark plutonics and gneiss, undated

Karamea Granite, Devonian

Fiordland crystalline rocks

Cambrian-Devonian trough sediments

Cambrian-Ordovician volcanics and sediments

Precambrian gneiss

0   100 km

N

Milford Sound

Karamea

Springs Junction

Alpine Fault

North-west Nelson and Fiordland were together until 100 million years ago or less.

Fig. 36. The oldest rocks (*after R. A. Cooper*).

22

CHAPTER TWO

Phase I:
The oldest rocks, in the west

THE beginning of the area that is now New Zealand took place on the sea floor somewhere off the coast of Gondwana about 600 million years ago. There, a row of volcanoes spewed lavas and ashes into the sea, building up volcanic islands (Fig. 37). After the volcanoes

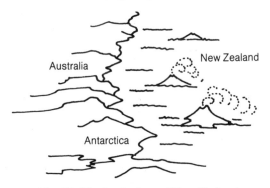

Fig. 37. The beginnings of New Zealand.

died down the sea floor was covered with sands and muds washed down from the land and with the skeletons of sea creatures. At times the sea floor buckled up and became land, only to be worn down or to sink again below the sea. Then the whole area was folded, altered and pushed up into a land which today is mountainous and mostly covered with bush. The relationships between the rocks are hard to see and much exploration is still going on, so what can be said now is very much a progress report.

The oldest rocks in New Zealand are near Charleston, on the West Coast — they are crystalline and known to be Precambrian. In the south, in Fiordland, most of the Cambrian and later sediments have been completely altered and recrystallised so it is difficult to work out

their original condition. In order to find sediments that still record the conditions under which they were deposited, to gain some idea of the geography of that far-distant time, we must go to North-west Nelson. Most of this chapter describes that area. However, we will look at the oldest rocks first.

THE CHARLESTON GNEISS

Like people, rocks can have two names. The last is the type of rock or family — for example, limestone, granite, gneiss (pronounced "nice"). The first name distinguishes particular members of a rock family — so Charleston Gneiss (with initial capital letters) is that gneiss found at Charleston and properly described and all the gneiss in the surrounding areas which is thought to have the same composition and origin.

Gneiss is a coarsely crystalline rock in which the minerals are separated into bands of light and dark minerals (Fig. 38). It is metamorphic

Fig. 38. Charleston Gneiss.

rock (from "meta", meaning altered, and "morph", meaning shape), a rock changed from its original character by great heat and pressure but without melting. The formation of gneiss actually takes place within solid rock. At different stages in the build-up of pressure and temperature, the various minerals in the original rock become unstable and their atoms regroup to grow into new minerals or groups of minerals. Eventually all the original minerals change into new minerals, forming gneiss.

Patches of an original mineral can sometimes be found inside zones of the new minerals if a microscope is used to view them (Fig. 39).

Fig. 40. Granite seam in gneiss, Little Beach.

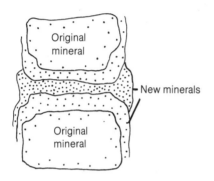

Fig. 39. New minerals forming from old in a metamorphic rock.

If the temperatures and pressures increase even further, parts of the newly formed gneiss may actually melt and move as a liquid into surrounding areas. The paler minerals within the gneiss, like quartz and feldspar, melt first and these may invade the surrounding gneiss as seams and masses of granite. (At this stage the distinction between metamorphic rocks and igneous rocks — from molten magma — becomes rather blurred.) Such a granite seam is shown in Fig. 40 in a rock from Little Beach at the mouth of the Nile River.

Charleston Gneiss forms part of the Paparoa Range, but it is most easily seen enclosing Joyce and Constant Bays (it was once called the Constant Gneiss), at Little Beach at the mouth of the Nile River, and also in roadcuts on the main Greymouth-Westport road. It is likely that these ancient rocks were part of the Gondwana landmass to the west, which provided the sediments that make up the rest of our oldest rocks.

Within the Charleston Gneiss are pegmatites,

which are seams of extremely coarse crystals which often contain the less common elements and minerals. In the gorse-covered stretch between Charleston and Constant Bay a pegmatite was once mined for mica; the outcrop still contains large crystals of muscovite mica, coarse quartz and feldspar and some opaque red garnet crystals (Fig. 41).

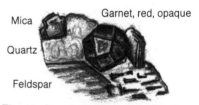

Fig. 41. Garnet in pegmatite, Charleston.

Inland, near Four Mile Stream, are pegmatites with quartz and feldspar crystals measuring a metre across and pale, greenish beryl crystals, so rare that unlucky collectors can dig all day without finding a single beryl even of the poorest quality (Fig. 42).

Fig. 42. Beryl on quartz, Charleston.

FIORDLAND

The old rocks of the Fiordland region are somewhat inaccessible, and detailed work on the geology of the area is still being done. In the Milford Sound area, perhaps the most often visited, the original rocks have been recrystallised to form gneiss, granodiorite (a rock between granite and diorite), diorite (Fig. 43) and gabbro. "The Chasm", signposted on the Cleddau River, on the road to Milford Sound, has been cut from fine-grained diorite. The boulders at Bowen Falls and the Homer Tunnel show some of the different textures of these plutonic rocks. Garnets and lime-green epidote are among the eye-catching minerals to be found.

Those who make the descent to the Manapouri powerhouse can see for themselves the chaotic nature of the Fiordland crystalline rocks. Sedimentary rocks containing Paleozoic fossils

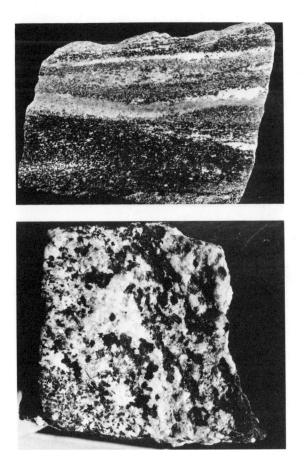

Fig. 43. Gneiss, Manapouri tunnel, and diorite, Milford Sound (bottom).

occur in very few places. The Ordovician layers in Preservation Inlet match those in North-west Nelson, and there are marbles in Kellard Sound like those of Takaka Hill, reminding us that these areas were once joined.

NORTH-WEST NELSON

Here the oldest rocks we can find are sediments containing a few fossils, but they are so folded we cannot tell their original positions. Clearly, there are three belts of rock running north/south: a central, mainly older sedimentary belt also containing volcanics; an eastern sedimentary belt with thick limestones; and a western sedimentary belt of sandstones. The original contact zones between the three belts have been wiped out by movement on the faults that now form the boundaries, as shown in Fig. 44. Some geologists argue that the central belt material was actually thrust forward from further south, while others believe the rocks are in their original positions but the central belt has been squeezed together and upwards.

While the debate continues, we can visit the area and enjoy the scenery and some rather special rocks. Its entire geological history can be traversed in the roadsides from Riwaka to the Cobb Valley. Going up the Takaka Hill from the east, the road first cuts through igneous rocks created in the great collision at the end of the first phase of our geological history, then it passes through sediments which have been altered beyond recognition to marble and schist. From Upper Takaka the road to the Cobb winds through sedimentary rocks formed in a period spanning some 150 million years, taking us on a journey back in time from the Silurian to the middle Cambrian Periods, the time of the central belt in the Cobb Valley.

Since these sediments still show the events of the past they can be sorted and described according to their ages. They show us a changing pattern of environments in the North-west Nelson area in ancient times. Geologists are slowly piecing together the sequence of events that gave us the rocks found in the area today.

The following diagrams are very simplified possible reconstructions of that ancient area. Their simplification is such that it could be likened to one picture being selected to show a typical New Zealand scene of the last 150 years.

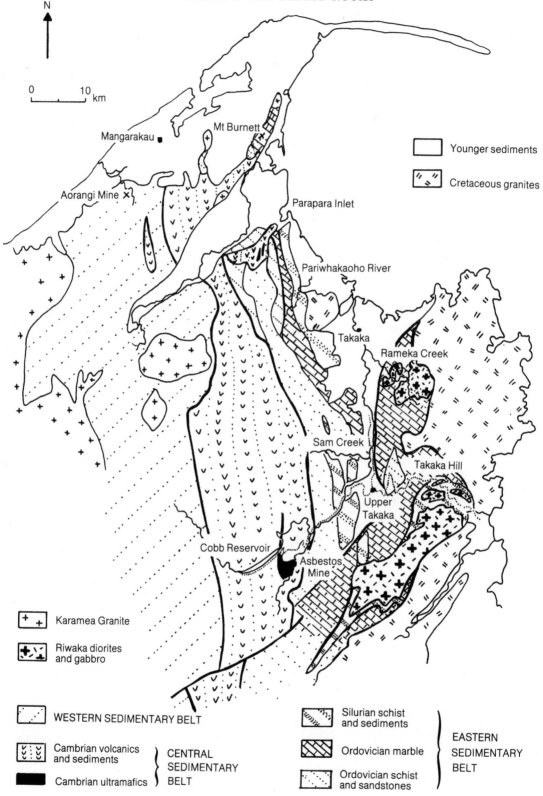

N

0   10 km

Mangarakau

Mt Burnett

Aorangi Mine ✗

Parapara Inlet

Pariwhakaoho River

Takaka

Rameka Creek

Sam Creek

Takaka Hill

Upper
Takaka

Cobb Reservoir

Asbestos
Mine

Younger sediments

Cretaceous granites

Karamea Granite

Riwaka diorites
and gabbro

WESTERN SEDIMENTARY BELT

Cambrian volcanics
and sediments

Cambrian ultramafics

CENTRAL
SEDIMENTARY
BELT

Silurian schist
and sediments

Ordovician marble

Ordovician schist
and sandstones

EASTERN
SEDIMENTARY
BELT

Fig. 44. North-west Nelson, map and cross-section (right) (*after R. A. Cooper.*)

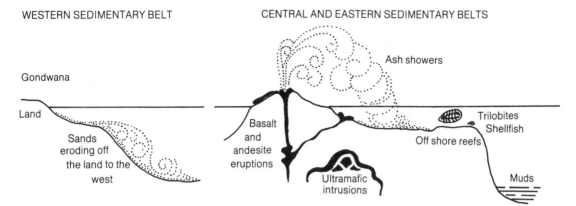

WESTERN SEDIMENTARY BELT

CENTRAL AND EASTERN SEDIMENTARY BELTS

Ash showers

Gondwana

Land

Sands eroding off the land to the west

Basalt and andesite eruptions

Trilobites
Shellfish

Off shore reefs

Ultramafic intrusions

Muds

Fig. 45. Events during the Cambrian Period in North-west Nelson.

The Cambrian Period: 600 to 475 million years ago

In the Cambrian Period the rocks of North-west Nelson's central belt were being formed off the coast of Gondwana, as shown in Fig. 45. Today these rocks can be seen in the Upper Takaka and Cobb Valleys along a narrow, winding road and well-marked forest tracks.

The volcanics here include andesites and basalts and debris from volcanic eruptions, but in general they are weathered and very hard to recognise.

The most distinctive of the Cambrian rocks are ultramafic igneous rocks, which perhaps formed deep under the ocean floor from sub-crustal material caught up in a collision zone below the volcanoes. The ultramafic rocks, rich in iron and magnesium have nearly all been altered to serpentine, a dark-green stone which sometimes has a greasy appearance on boulder edges. A mineral that often occurs with serpentine is talc — a soft, pale mineral which is crushed to make the familiar talcum powder. A mass of talc makes up the rock called soapstone,

which is recognisable by its greasy texture. Soapstone is a good carving material because it can be cut with a knife.

The biggest deposit of ultramafic rocks is above the Takaka River, several kilometres beyond the powerhouse. There is a signposted track, the Asbestos Forest Walk, which can be walked in somewhat more than an hour each way to and from the old mine where asbestos was quarried between 1949 and 1963 when the venture became uneconomical. Collectors still find serpentine with veins of short cross-fibres and sheared masses of longer fibres (Plate 1C). Blocks of carving soapstone can be found, and rarer pink thulite is present in patches in the serpentine on the dumps.

On the road down to the Cobb Reservoir, part-way down the first leg of the zigzag, there is another green outcrop of serpentine and soapstone. Beside these rocks are some of the few recognisable volcanic rocks — fine-grained felsic (pale) lavas.

Across the dam, to the north, another ultramafic area is being mined for talc and magnesite, which is used in fertilisers, pottery and

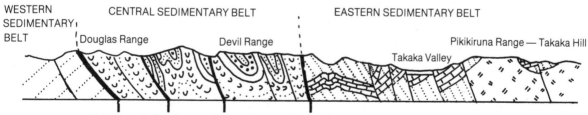

WESTERN SEDIMENTARY BELT

CENTRAL SEDIMENTARY BELT

Douglas Range

Devil Range

EASTERN SEDIMENTARY BELT

Pikikiruna Range — Takaka Hill

Takaka Valley

Main thrust faults where rocks have been pushed against others for great distances, either along, or sideways and upwards papermaking. The magnesite, or magnesium carbonate, found here has cream-coloured cleavage surfaces. Some of the small lumps of talc are stained bright-green with chromium, an element often found in the ultramafics.

At the head of the lake are patches of sedimentary rocks, fine-grained mudstones and limestones containing the oldest of New Zealand's fossils that can be seen without a microscope.

Perhaps we should now stop and investigate sedimentary rocks. "Sedeo" is a Latin word meaning "I sit"; sedimentary rocks are rocks made from particles that first moved along, then stopped to rest. Once uplifted, rocks begin to crumble and are carried down from the land until they reach the depths of the ocean, as illustrated in the drawing of the rock cycle (Fig. 3) on page 5. Most of us have seen rocks on this journey.

In mountain screes the stony slopes are built of rough pieces of stone which have been freshly broken from the parent rock. When they reach the rivers they are tumbled together and smoothed as the river sorts them into beds of boulders and gravel with margins of sand and flood backwaters with finer muds. Eventually the river empties its debris into the sea, where the heaviest material is first dropped by the currents. Finer muds are carried out further before being dropped, but eventually even the smallest particles settle on the ocean floor (Fig. 46).

In central areas of oceans there is no material from the land; only minute shells drop down from drifts of tiny plants and animals known as plankton. Sometimes chemicals crystallise from the sea water; the manganese that forms small lumps or nodules on the Chatham Rise is an example.

Many shellfish live in the shallow seas and leave their shells in the sediment on the sea floor. Shells are formed of lime (calcium carbonate). When more than half a rock is made of lime from this source it is called a limestone. Limestone may contain whole shells, pieces of shell or lime dissolved and recrystallised. Large shells in the rock are easy to see, but a microscope is needed to view the tiny shells that have accumulated over millions of years as an ooze on the floors of the deepest oceans.

With the passing of time sediments are

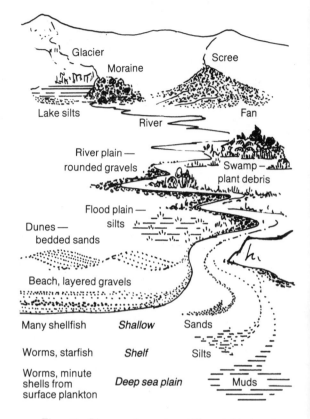

Fig. 46. Places where the different kinds of sedimentary rocks may form.

cemented together and harden. This can happen at any stage in their downward journey — while they are still screes, while they are river sands or deepest oceanic oozes. All may be preserved as rocks, showing the environments in which they once existed and which formed them. If you look closely at any sedimentary rock you may see structure typical of the place where the rock was formed. Some special sediments will be discussed later; but just the size of rock fragments can tell you the distance they have travelled, their material can reveal the type of land that was their source and fossils may tell you whether the sediments were deposited on land or under the sea.

The appearance of the sediments at the head of the Cobb Reservoir reveals that they were formed in two very different environments. Some are fine muds and silts, which would have built up in relatively deep water; but among them are small, lens-shaped patches of grey

Fig. 47. Trilobite rock, Cobb Valley.

limestone (Fig. 47) which once were shell banks in a shallow sea. The dominant shells are fragments of the external skeletons of trilobites, which are extinct marine animals related to crabs and woodlice. (Their name, trilobite, is derived from the three lobes of the body, Fig. 48, which run north/south like the east, west and central belts of North-west Nelson).

In the Cobb limestone there are at least 16 kinds of trilobite fossils, which once lived together in well-defined communities. On a world scale, trilobites were common from the Cambrian to the Devonian Periods but died out altogether at the end of the Permian. They are very useful time indicators and show that the Cobb limestone is some 550 million years old.

Like crabs, trilobites moulted as they grew; a complete skeleton is rare, but moulted pieces once littered the sea bed. Since the heads and tails were fused together into single plates, these parts tended to survive when the jointed segments of the body fell apart. Waves swept the shed trilobite pieces with other minute shells into banks which were later cemented together,

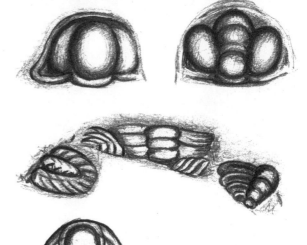

Eye

Fig. 48. Trilobites. Left, whole trilobite. Top, moulted heads and tails found in the Cobb limestone.

29

then they slumped into nearby deep water where mudstones had been lying. This is how the grey limestone came to be combined with the muds and silts of the Cobb area.

Two other distinctive types of rock can be seen after a 10-minute walk up the road-track to Lake Sylvester on the western side of the Cobb Reservoir. In the first deep-cut creek are large boulders of conglomerate (Fig. 49), a sedimentary rock formed of rounded pebbles cemented together. These boulders have descended from an enormous deposit topping the ranges which runs for many kilometres and is some 450 metres thick. The deposit may have formed as a bouldery flood-plain around a mountain range. The many different kinds of rock in the pebbles are those that must have been part of that Cambrian mountain range. The banks of the stream consist of bands of metamorphic grey slate with seams of tan-coloured ankerite, an iron-magnesium carbonate, in sandstones.

Fig. 49. Boulders of conglomerate and banks of slate and sandstone, Cobb Reservoir.

These rocks are just some of those that make up the oldest group of rocks in North-west Nelson. Geologists are still hiking from the Cobb Valley into the surrounding wilderness to find and explain all the rocks of the Cambrian Period. Further information may be found in the *Geological Society Guidebook to the Cobb Valley* by Roger Cooper.

The Ordovician Period: 475 to 425 million years ago

Most of the rocks of the western and eastern sedimentary belts were formed during the Ordovician Period (Fig. 50). The two belts represent two very different environments. The rocks in the eastern belt were produced mainly by lime-secreting animals living in warm, shallow seas where very little land material arrived until late in the Ordovician. The rocks in the western area were formed in relatively deep water with sediments coming off a continental land mass; they are mainly fine-grained mudstones and shales; the latter are mudstones made up of platy clay minerals that can be split into layers.

In 1874 gold was found in quartz seams cutting the shales of the western belt. Today the mines are abandoned and the batteries rusting, but geologists still visit the area because the shales near the Aorangi Mine (approximately 8 kilometres south-west of Mangarakau) contain a good series of graptolite fossils (see Fig. 51) which link our land with the rocks bearing an identical series in Victoria, Australia. It is a 1½-hour walk to the fossils from the end of a rough farm road, and permission to go there must be obtained from the farm manager at Mangarakau. Graptolite fossils can be found also in Chaffey Stream, 6 kilometres along the track from the head of the Cobb Reservoir, but

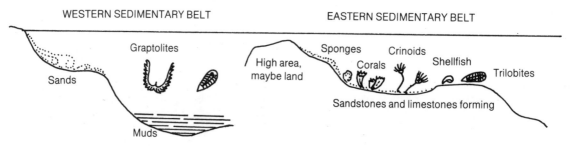

Fig. 50. Events during the Ordovician Period in North-west Nelson.

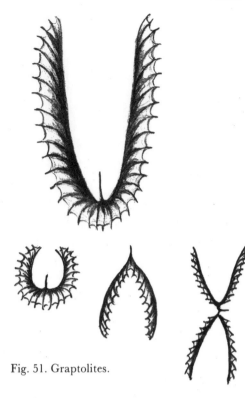

Fig. 51. Graptolites.

Fig. 52. Graptolites on slate, Aorangi Mine.

nowhere are they readily accessible. Graptolites (Figs. 51 and 52) were primitive marine animals which formed colonies of polyp-like creatures linked along a single connecting chord. Each lived in a separate cup on a branching stem. They evolved rapidly, changing their pattern of branching, and because each form was short-lived they provide a good means of distinguishing different layers of rock.

At the Aorangi Mine graptolites form white impressions on the black shale, sometimes spangled with metallic iron pyrites (see page 38); but on paler mudstones they can appear simply as rusty marks and are very easily missed.

Further south, near Reefton, the sandstones and mudstones of the Waiuta area were assumed to be Precambrian because no fossils had been seen, but in 1974 a single graptolite species was found and these rocks are now considered as Ordovician and part of the western sedimentary area. At Reefton the rocks are also gold bearing, and ruined mines can be visited. A narrow road (signposted 21 kilometres south-east of Reefton on State Highway 7) leads

to the Waiuta-Blackwater Mine, which closed down in 1951 when the mine flooded. The poppet head can be seen on the plateau above the remains of Waiuta township, and the impressive remains of the cyanide tanks are visible below on the banks of the Snowy River. Nearly all the Reefton mines were dug in quartz veins in the Ordovician rocks. The Forest Service maintains tracks to the mines in the Blacks Point area and issues a pamphlet on walks in the Murray Creek goldfield.

The same sediments can be seen along the coast road north of Greymouth between Thirteen Mile Creek and Barrytown (Fig. 53). To date no one has seen fossils in these layers of sandstone and mudstone.

Fig. 53. Sandstones south of Barrytown, thought to be Ordovician in age.

31

The fossils that have been used to date the rocks in the eastern area have been found mostly on mountainsides far from any road. The sediments in the more accessible areas have been altered by heat and pressure and most of the fossils have been destroyed. The few that remain are very hard to identify. However, some fossils may be found without wandering too far. On the way into the Cobb Valley there are great thicknesses of sandstone, siltstones and mud-stones. On the bluff opposite Barrons Creek (Fig. 54), where the entrance to the forest park is signposted, there are a few small, rusty corals (Fig. 55) hiding in the mudstone, and one complete trilobite has been found. The Geological Survey would welcome more fossil specimens, especially graptolites.

On the Ordovician sea bed there must have been a series of enormous shell banks which shifted in position over the millions of years during which they existed. Most of these banks have been altered to marble and recrystallised, so any shells or fossils remaining have very blurred outlines. However, the marbles on the west side of the Takaka Valley are a little less altered, and sponges and corals can be found weathering out of the boulders up Sam Creek. This is the first river bed on the Uruwhenua Road which turns south from the main road just below Lindsays Bridge on the Takaka River. (There is another Sam Creek up the Takaka Gorge.)

The sponges appear as rough, brownish lumps on weathered surfaces and cannot be seen on a fresh, broken face. The corals are whiter and show the structure of their coral walls a little more clearly than the sponges show their shape (Fig. 56).

Fig. 54. Roadcut on the bluff opposite Barrons Creek.

Fig. 55. Coral from Ordovician mudstone at the Barrons Creek roadcut.

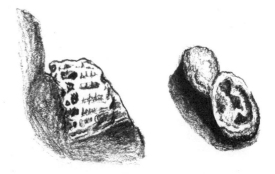

Fig. 56. Ordovician coral (left) and sponge (right) from Sam Creek.

WESTERN SEDIMENTARY BELT   EASTERN SEDIMENTARY BELT

Some folding and uplift   Sands   Shellfish

Fig. 57. Events during the Silurian Period in North-west Nelson.

The Silurian Period: 425 to 413 million years ago

There seems to have been some disturbance in the west at this time because the only sediments building up on the sea floor were in the east (Fig. 57).

At Hailes Knob, high in the mountains to the west of Motueka, a few Silurian fossils — shellfish and trilobites — have been found. These are in quartz sandstones; similar sandstones form ridges beside the road to the Cobb Valley a few kilometres from the Upper Takaka turnoff before entering the gorge. There is a boulder by the roadside beside the Silurian ridges which shows folds of marble and sandstone (Fig. 58). It is possible to see a few tiny shellfish and crinoid (sea lily) stems in this rock.

Crinoids (Fig. 59) are related to brittle stars and starfish. Their skeletons are made up of plates of calcite. When the animal dies the plates fall apart. Complete tops are rarely found, but the distinctive round plates of the stems are fairly common in some marbles and also in later Devonian sediments.

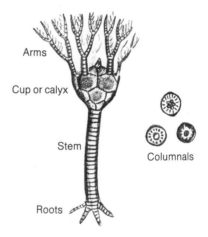

Fig. 59. Crinoid or sea lily.

Fig. 58. Silurian rocks before the entrance to the Takaka Gorge. Ridges show in the pines to the east of the road and the boulder stands in bracken on the other side of the road.

33

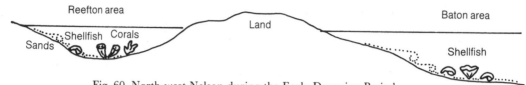

Fig. 60. North-west Nelson during the Early Devonian Period.

The Devonian Period: 413 to 355 million years ago

At first sediments continued to build up only on the eastern side of North-west Nelson, but after the disturbances in the west a shallow sea covered the area and for a while a coral reef grew along the coast (Fig. 60). Then, about the middle of the Devonian Period, the entire long cycle came to an end as the crust buckled and the rocks were lifted high.

Today there are only two small areas of Devonian rocks to be found in New Zealand. An eastern area is well up the Baton River, south of Motueka, where sandstones contain various shellfish fossils. The western area is much more accessible. Approximately 6.5 kilometres south-east of Reefton, and 1.4 kilometres from Lankey Creek, the main road to Lewis Pass cuts through a dark limestone bluff beside the Inangahua River, and corals can be seen in the limestone. A few metres further along the road the rocks change to steeply dipping layers of finer sands and siltstones, with remains of shellfish and other fossils (Plate 1M). Another accessible area of these rocks is up Lankey Creek, which is signposted on the main road (SH 7) about 4.9 kilometres east of Reefton. Here the Forest Service has a posted track up to the gold batteries and fossil site, but fossils can also be found in boulders in the creek down by the road.

The limestones at Reefton contain vast masses of corals and other creatures which were built up into a reef. (Corals [Fig. 61] first developed in the Ordovician Period. The small, solitary corals found by the Takaka River, mentioned on page 32, are typical of those found in deep, soft muds. Those found at Reefton formed part of a coral reef and grew very much larger, or in branching colonies.) The corals are evident as pale outlines in the dark limestone (Fig. 62) and are especially noticeable on the weathered surfaces of rocks in the Inangahua River below the road.

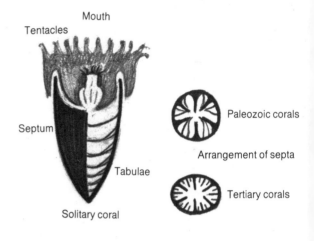

Fig. 61. Coral structure.

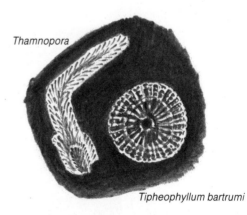

Fig. 62. Devonian fossil corals, Inangahua River.

34

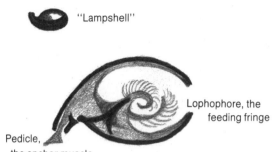

"Lampshell"

Lophophore, the
feeding fringe

Pedicle,
the anchor muscle

Fig. 63. Brachiopod structure. Even on a cast the prominent hook of the pedicle opening will often distinguish a brachiopod from a bivalve mollusc like a cockle.

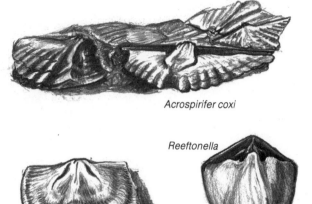

Acrospirifer coxi

Reeftonella

Reeftonia and columnal

Fig. 64. Devonian brachiopods, Reefton.

In the browner sandstones and siltstones of this area there are many fossilised shellfish, including large and clear remains of lampshells or brachiopods. Brachiopods are not related to common shellfish like cockles or mussels, which belong to the mollusc group; they are a completely different kind of sea animal. Each shell is symmetrical, but the two shells differ — in profile they resemble the ancient oil lamp, hence their common name. Whereas molluscs fill most of the space between their two shells, brachiopod bodies are very small and most of the cavity between the shells is taken up by a fine loop of shell supporting a fleshy fringe which waves food particles into the mouth. See Fig. 63.

Brachiopods are one of the oldest groups of animals in the world. They were once so widespread and diverse that they have proved good marker fossils, but today they are relatively uncommon. New Zealand is one of the few places in the world where they have formed large colonies (especially in the south, in Foveaux Strait and Fiordland). Our oldest brachiopods are Cambrian, but these can be seen only under a microscope; the Silurian brachiopods are present in a very remote area. The Devonian brachiopods near Reefton are the oldest likely to be seen by travellers. Their shells have been dissolved by ages of percolating ground waters, leaving only their impressions, which have sometimes been distorted during the folding of the rocks (Fig. 64). Other fossils at

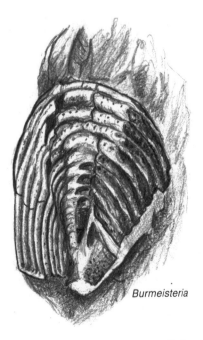

Burmeisteria

Fig. 65. Devonian trilobite tail, Reefton.

Reefton include true molluscs, rare trilobite remains and many pieces of crinoid stems (Fig. 65). All these fossils are typical of life in a fairly warm and shallow sea.

35

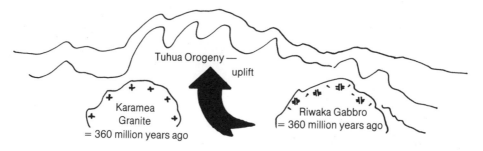

Fig. 66. Events during the Middle to Late Devonian in North-west Nelson.

THE TUHUA OROGENY — THE FIRST UPHEAVAL (LATE DEVONIAN)

The long period of sediment build-up ended with a period of great pressure and uplift, called an orogeny, when all the sea floor sediments were pushed up, folded and welded on to the old land mass to make a new mountain range beside the sea (Fig. 66).

In the 200 million years it took to produce the rocks so far described there had been various upward movements in the area. We know this because in some places deposits were interrupted, then pebbles of previously formed material began to appear in the sediments. This could have happened only after the earlier rocks had been lifted up to the earth's surface and exposed to the forces of erosion. However, in the middle of the Devonian Period all deposition in this part of the world came to an end. All the rocks that had been forming were forced together into the steep folds seen in the hills today. They were lifted high and many were greatly altered by the intense pressure that folded them.

When we looked at gneiss we saw how minerals under great heat and pressure may break down allowing the atoms to regroup into new minerals. In the same way during the Tuhua Orogeny whole rocks were reconstituted and new groups of minerals were built up slowly, atom by atom. Sandstones and mudstones became schist — a rock containing parallel layers of minerals, especially mica, which gives the rock its shimmering, silvery surface. Schist itself has cleavage, therefore outcrops of schist have flattened, shiny surfaces. Some of the Cambrian sediments with a high proportion of mafic (iron-magnesium) volcanic debris became greenschists, coloured with the dark-green, mica-like mineral known as chlorite. They can be seen on the roadside downstream from the Cobb Hydro Station. They also contain layers of white quartz, creamy feldspar and pale mica, and all the rock is studded with small crystals of pyrite, the brassy iron sulphide.

Some of the Cambrian ultramafics altered to form a very dark-green schist composed almost entirely of chlorite. A patch of this greenschist on the Parapara River above Richmond Flat contains sprays of black tourmaline crystals (Fig. 67) and extremely rare masses of dull blue corundum or sapphire (Plate 1D). It is sited a two-hour walk from the Rockville Caves. Visitors in 1981 had to fight their way through head-high gorse at the last hill, where they found far more prickles than tourmaline.

Fig. 67. Black tourmaline in green chlorite schist, Richmond Flat.

The tourmaline group is made up of very complex boron silicates which form long crystals; their colour depends on their chemical composition. Schorl, a black, iron-rich tourmaline, can be distinguished from the more common black hornblende because tourmaline has no cleavage and is triangular in cross-section; the surface is often striated along the crystal. Tourmaline is rare in New Zealand, but black schorl has been found in the granites of the Buller Gorge, near Lyell, and in some other granites. Small brown and green tourmalines also have been found in the quarry at the turnoff to Parapara Beach, Golden Bay, where the material was excavated for the causeway across the inlet.

This quarry is situated in one of the bands of schist that formed from the different layers of sandstones found in the eastern sedimentary belt. Because of the degree of their alteration it has been difficult to link the sandstones with their schist equivalents, but the Onekaka Schist (named after the area), which runs south from this quarry, is now thought to have formed from the youngest Silurian sandstones.

Whatever its age, the schist at Parapara Quarry contains flakes of a green-coloured chrome mica known as fuchsite, which is coarsest in the white quartz on the seaward side of the quarry. Here there are also 1 millimetre green tourmalines and 2-3 millimetre crystals of red rutile, a titanium oxide. On the floor of the quarry is a seam of rotten or weathered crystals of bluish kyanite, an aluminium silicate which forms blades in rocks that have been altered under pressure as well as temperature. It tends to crumble when it is collected.

Better kyanite (Plate 1J) has been found in the Pariwhakaoho River some 10 kilometres west of Takaka and shown on the map, Fig. 68.

Fig. 68. Map of the Pariwhakaoho area, Golden Bay (*after Grindley*).

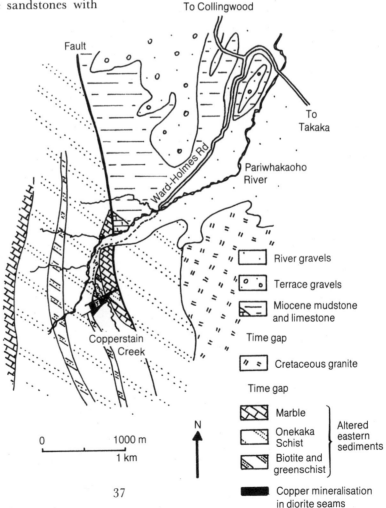

A forest park track to Parapara Peak leads from the end of Ward-Holmes Road and enters the bush below Copperstain Creek. The boulders in the river from here back to the road end contain samples of all the rocks in the steep and bush-clad hills, and collectors may gather the minerals and fossils with no greater cost than wet feet and sore calf muscles from boulder-hopping.

The blue kyanite is found in boulders with white quartz and a schist mainly composed of coarse, black mica (biotite), which has come from two narrow belts of rock hidden on the hillsides (see map). Most of the western bank of the river is composed of Onekaka Schist, a silvery-grey, wrinkled (crenulated) schist common in the river.

Greenschist here and up Copperstain Creek contains pyrite (iron sulphide, or fool's gold) and chalcopyrite, the copper-iron sulphide. Both these sulphides are metallic and brassy; but iron sulphide (pyrite) is harder than steel and much paler in colour than the rich, brassy, coppery chalcopyrite, which can be cut with a knife and rarely forms in good crystals. Pyrite (Fig. 69) does form crystals of various shapes — cubes or 12-sided pyritohedrons or, less commonly, octohedrons or combinations of these forms. The surfaces of pyrite crystals are often striated and they can tarnish.

Another mineral occasionally found in the greenschist is scapolite, which occurs as white, woody-looking crystals which are square in cross-section and have a silky cleavage. (Fig. 69). This sodium-calcium silicate is generally found as microscopic grains or in massive seams, but some crystals have been found.

Chalcopyrite is more common in Copperstain

Creek, where thin fingers of granodiorite poked into the schists and marbles from the later granite formed just to the east in the Cretaceous Period. A little way up the creek from the track was an area of rusty rock known as a gossan, where an ore deposit came to the surface and was leached by the rains until only a rusty, corroded area remained. Prospecting has shown that mining of the ore would probably not be economic, but collectors have found crusts of the green copper carbonate malachite, and the turquoise-blue copper silicate, chrysocolla, in the creek and in a hole up the bank.

Also found in the Pariwhakaoho River bed is a hardened black mudstone containing garnet crystals (Fig. 69). The crystals are well formed and up to 8 millimetres across, but they are not of gem quality (that is, not clear, well coloured and flawless crystals that could be cut into gemstones). The mudstones were cooked when the Cretaceous granites came into the surrounding area as a molten mass then cooled and lost their heat to the surrounding rocks. The mud reformed into fine crystals and hardened, and garnets grew from the iron, aluminium and silica in the mud, but there was little pressure to squash the growing minerals and produce the layering in the rocks characteristic of schists. The garnets found here have inclusions and cracks which is why they cannot be cut into gems.

Looking upstream (Fig. 70) we can see the lower left-hand ridge, which is the strip of marble shown in Fig. 68. In the distance above the centre is Parapara Peak, which is off the map and made up of much younger sandstones and conglomerates from the Permian Period. These rocks really belong in the next chapter, but

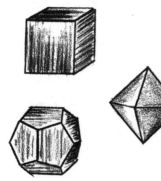  

Fig. 69. Minerals from the Pariwhakaoho River. Left, idealised pyrite crystals. Middle, scapolite in schist. Right, garnets in mudstone.

Fig. 70. Pariwhakaoho River, looking upstream.

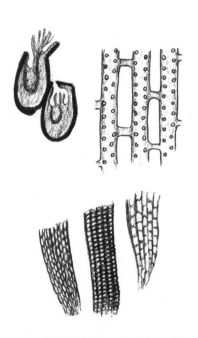

Fig. 71. Permian *Fenestella* from Parapara Peak.

because the Parapara Peak rocks are unique (they are the only Permian rocks in the country that rest on older basement and contain certain fossils), and because we are up this particular creek, or river, they have been included below.

Parapara Peak rocks

After the Tuhua Orogeny, when all the land of the first phase was welded together, most of this area wore away and was covered by a shallow sea. Here, in middle Permian times, pebbles of schist and sandstone were washed down rivers in flood to form conglomerates on the sea floor, and shellfish flourished in the shallow waters. Today these rocks are high on Parapara Peak, but a few boulders containing casts of shellfish

and bryozoans may be found in the Pariwha-kaoho River. (Bryozoans are one-celled animals which live in colonies; their fossils are sometimes mistaken for corals. One bryozoan, *Fenestella* (Fig. 71), produced a lacy, netlike skeleton in which the animals lived like tenants in multi-storeyed flats.)

Some of these conglomerates and sandstones were also later altered by the neighbouring granites of the Cretaceous uplift at the end of the second major phase of our geological history. Some of them contain small garnets within the pebbles of the conglomerate as well as in the sandstone around the pebbles, showing that the entire rock has been heated.

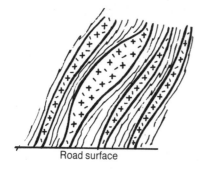

Road surface

Fig. 72. Lit-par-lit interfingering of granite and schist, Pupu Springs road junction by Takaka River bridge.

Fig. 73. Sedimentary layering and bands of quartz in marble, opposite the Log Cabin on Takaka Hill.

The Onekaka Schist reappears in the roadcut at the turnoff to Pupu Springs (2 kilometres north of Takaka on SH60 to Collingwood and well signposted). Here the rock has been intruded by granite fingers which have pushed their way between the layers of schist. The alternating layers of granite and schist are becoming harder to see as the cutting weathers, but a careful, close examination will reveal thick layers of coarsely crystalline granite and beds of flattened, grey, layered schist. Some of the granite layers are almost a metre thick. The interfingering intrusion of granite is known as lit-par-lit, or bed-by-bed (Fig. 72).

A similar schist, but one made from older Ordovician sandstones, is visible where the road begins its steep drop into the valley from the summit of Takaka Hill as you leave the great marble plateau, the most showy of the altered rocks.

Marble is formed from limestone when all the original calcite in the rock has been recrystallised into coarse grains and no trace of the original sediment remains. In some places in this area sedimentary structures do show through: on the roadside west of the parking place for the Ngarua Caves some layers of sediment can be seen, now folded, containing lines of sandy grains like the rock in Fig. 73. Sometimes the impurities in the sediments become new minerals — crystals of pyrite or garnet.

The white, grey and green marbles of this area have been quarried as building stone, most recently used for the Beehive. They are also used in the making of agricultural lime, as a flux in glassmaking and for ornamental chips in aggregates. The marble to the west of Collingwood at Mount Burnett contains so much magnesium that it is dolomite, which is also quarried for fertiliser and for use in glassmaking. This very attractive rock occurs in rich shades of cream, gold, pink and maroon.

Marble, like limestone, dissolves in rainwater, given geologic time. The surfaces of the rocks on Takaka Hill are furrowed, ridged and crevassed (Fig. 74). and instead of running into rivers the rainwater that has fallen on this area has eaten its way into sinkholes, which in places merge into cave systems. At the top of the hill the Ngarua Caves are open to the public, and Harwood's Hole in Canaan Valley is some 300 metres deep. An even deeper group of caves in the area is still being explored. They are all the work of rainwater.

Rain seeps into the soil and soaks down until it reaches a zone where all the cracks and pores in the rock are permanently full of water — beneath the water table (Fig. 75). Rainwater absorbs some of the carbon dioxide from the air and more from the soil and decaying vegetation, forming a weak carbonic acid solution. This dissolves the calcite in the limestone and marble and excavates passageways along former joints in the rock as it flows towards sea level, reaching the surface of the ground in springs, such as Pupu Springs near Takaka.

If the water table is lowered as river valleys deepen, or if the land is lifted, air enters the passages. Then the seeping water and dissolved calcite loses some of its carbon dioxide to the air and the water becomes less acidic, so the calcite is left behind and crystallises on the surface of the cave. In time a tube or stalactite builds up, and a matching mound or stalagmite forms where the water splashes down on the floor of the cave (Fig. 76). Eventually wonderful traceries are built up. The process takes millions of years and is an irreplaceable work of water. Collectors should remember cave decorations must therefore never be damaged. Similar formations can sometimes be found in limestone quarries where they may be collected quite legitimately.

Fig. 74. Waterworn marble on Takaka Hill.

Fig. 76. Cave formations.

Fig. 75. Erosion of caves.

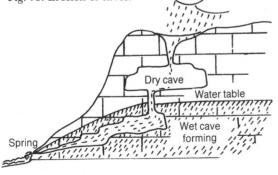

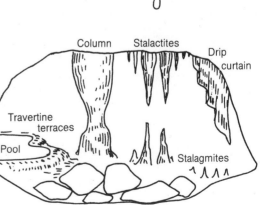

Granites in the orogeny

During any great disturbance and uplift the rocks at the bottom of a pile may melt, producing enormous masses of granite or more basic plutonic rock. In the Tuhua Orogeny the Karamea Granites pushed their way into the western area of North-west Nelson and in the east the Riwaka group of rocks, more mafic in content, was formed.

Karamea Granite (Fig. 77) is a coarse, crystalline rock, often containing large, pink crystals of orthoclase feldspar; it is in the Buller Gorge as well as at Karamea. In rare localities along the Karamea and Glenroy Rivers the crystals have formed in well-defined rings around some nucleus to make a most unusual and attractive rock known as orbicular granite (because it seems to be made of orbs or balls). See Fig. 78.

The Riwaka igneous group is made up mainly of diorites and gabbros. The biggest outcrop of these rocks is on the east of the Takaha Hill, from the Riwaka Valley south past the Graham Valley, where the rocks contain nickel-iron ores. The Riwaka mafics have pushed their way into the Onekaka Schist, fingering into the layers on the edge of the schist in a large-scale version of the layering at the Pupu turnoff (Fig. 79). Beds of Onekaka Schist among the Riwaka diorite can be seen on the roadside ascending Takaka Hill from Riwaka.

Another part of the Riwaka Group is exposed in Rameka Creek, east of Takaka. Here the diorites are made mostly of laths of black hornblende and white feldspar, with some pale-green epidote and dark-green chlorite, both

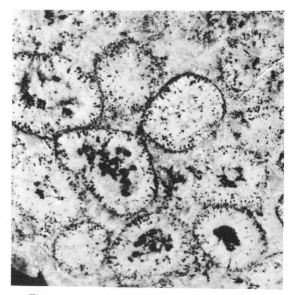

Fig. 78. Orbicular granite from Karamea: the orbs measure about 50 mm across.

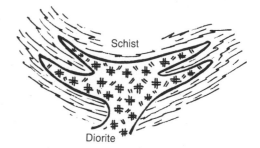

Fig. 79. Interfingering diorite and schist on the Takaka Hill, where the road cuts across the layers.

Fig. 77. Karamea Granite at Tauranga Bay, south of Westport.

introduced when the rocks were heated again when the main mass of the Cretaceous granites was cooling to the east. In places the rocks are cut by bands of much coarser crystals 20-25 centimetres long. Cavities in the rock may contain unusual minerals such as the pale-green, wedge-shaped sphene (Plate 1K), a calcium-sodium silicate which may be seen in the rocks on the last corner before the ford to the Rameka Track.

Granite and other plutonic rocks often weather in a peculiar fashion, with thin sheets of rock splitting off as the surface expands and contracts during daily temperature variations. A diorite boulder up Rameka Creek (Fig. 80) shows such "onion-skin" weathering.

When the hot masses of the Riwaka igneous rocks pushed into the limestone marble a zone of new rock was formed by a process known as contact metamorphism. One such contact zone or skarn can be seen in a bank on the Rameka Track about 100 metres from the ford (Plate 1L). On contact with the marbles the original diorite gave out not only heat but also liquids which moved out into the surrounding marbles, taking with them various elements from the diorite. Some elements travelled further than others, the least mobile remaining in a narrow belt close to their source, as shown below.

All the invading elements joined with the calcium in the marble to make various new minerals arranged in bands according to the distances travelled by their component elements (Fig. 81).

In the skarn on the Rameka Track the crystals are very small and their zoning is shown in the colouration of the rock: the skarn is very narrow, measuring only about 0.5 metres. Loose boulders in the creek and on the roadside up to the ford may be found containing larger and more colourful examples of the minerals within the skarn. Diopside and grossular (a member of

Fig. 80. "Onion-skin" weathering on diorite, Rameka Creek.

the garnet group of minerals) are both pale green, but the grossular found here is a little more translucent and has no cleavage, while the diopside has good cleavage because it is a member of the pyroxene group. Andradite, another garnet, is a deep purple-brown colour with rare crystal faces in these outcrops, while the epidote is a bright yellow-green or chartreuse. The final mineral, wollastonite, grows in silky, white, fibrous crystals which are not common in Rameka Creek but have been found in the Riwaka River and Holyoake Valley.

The alteration that formed all these minerals was part of the Tuhua Orogeny.

By the end of this first great disturbance in New Zealand's geological history all the layers of sediments and volcanic debris that form the foundations of North-west Nelson and Fiordland had been uplifted and made into land. Intense pressure from the east had crumpled and squeezed the layers into folds that run north/ south (see Fig. 44 on page 27) and bodies of granite and diorite had formed within the new land mass.

The land was ready for the next cycle.

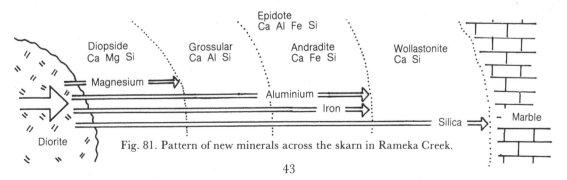

Fig. 81. Pattern of new minerals across the skarn in Rameka Creek.

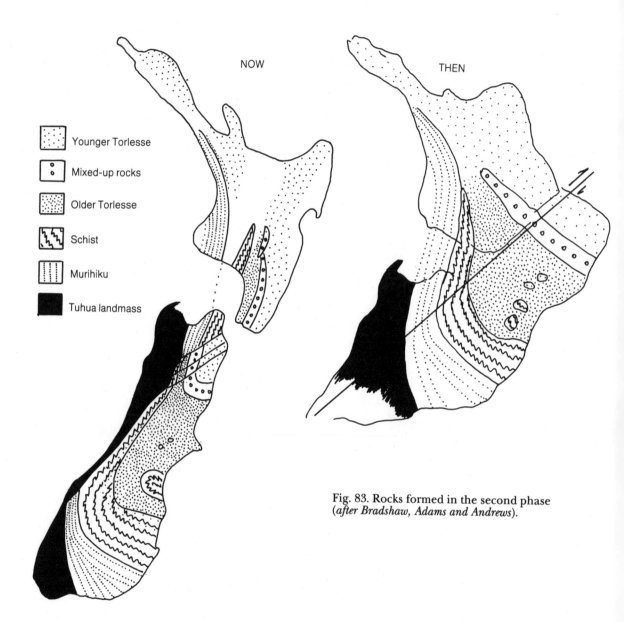

NOW

THEN

Younger Torlesse

Mixed-up rocks

Older Torlesse

Schist

Murihiku

Tuhua landmass

Fig. 83. Rocks formed in the second phase
(*after Bradshaw, Adams and Andrews*).

CHAPTER THREE

Phase II:
The eastern block

AFTER the upheaval of the Tuhua Orogeny, during which the first strip of our land was welded on to the Gondwana continent, a new period of growth occurred. Once more volcanoes erupted, the land was eroded and sediments built up on the ocean floor (Fig. 82). The rocks

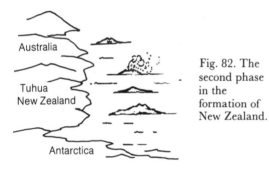

Fig. 82. The second phase in the formation of New Zealand.

of this second cycle were the foundations for the rest of New Zealand. They can be divided roughly into two groups, with a fault or belt of metamorphic rocks between them (Fig. 83). Both groups were formed before movement along the Alpine Fault wrenched the rocks of Nelson and Otago some 500 kilometres apart.

The Torlesse rocks on the east (named from the range in Canterbury crossed at Porters Pass on the way to Arthur's Pass) are mostly the familiar greywacke of the main ranges, built up of strongly folded layers of hard sandstones and mudstones rich in quartz, such as that which would have eroded off a land with a granite base. Fossils are few and far between. In contrast, the Murihiku rocks to the south and west are gently folded, with a good series of fossils buried in sediments rich in dark, volcanic

debris. Murihiku is the Maori name for Southland, where these rocks show up particularly well.

Because the Murihiku rocks provide a fairly clear picture of what was happening in their area during the second phase of New Zealand's geological history — from the early Permian some 265 million years ago to the end of the Jurassic Period some 130 millions later — we will look at them first.

THE MURIHIKU ROCKS

These rocks probably formed in a single, straight belt which has since been twisted to an S-bend and folded. Such large-scale folds in rocks are called synclines and anticlines (Fig. 84). When the tops of these folds are planed off

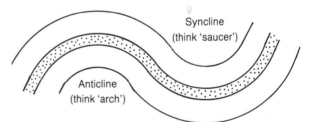

Fig. 84. Folding in rocks.

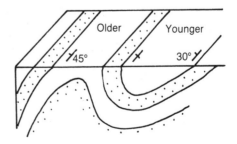

45

by erosion, anticlines reveal older rocks in their centres, whereas the central rocks of synclines are younger than the rocks on each side. On geological maps the dipping layers are shown by the symbol ∠, the long line being parallel with the edges of the layers and the short line pointing down the slope as shown in the illustration. The figures refer to the angles of the slopes from the horizontal, showing whether the layers are tilted steeply or gently sloping.

In Southland, the layers of the Murihiku rocks are tilted up on edge like stacked corrugated iron, forming the Southland Syncline (Fig. 85). The edges of the layers form ridges along the hills and can be followed from the Takitimu Mountains to the sea, especially from the air, although they are easily seen from the road between Mossburn and Gore. In Nelson the eastern flank of the same syncline outcrops along the Richmond Hills, but the western flank is covered with thick river gravels. In South Auckland the whole syncline can be seen and is known as the Kawhia Syncline.

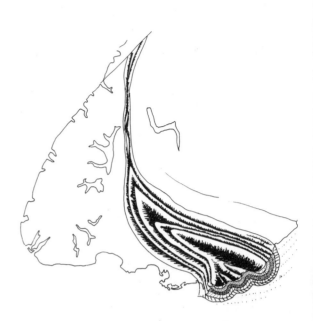

Fig. 85. The Southland Syncline, showing the edges of the folded layers. The change in dip from steep northern ridges to gentle southern slopes can be seen in the core of the syncline looking east in the Hokonui Hills (below).

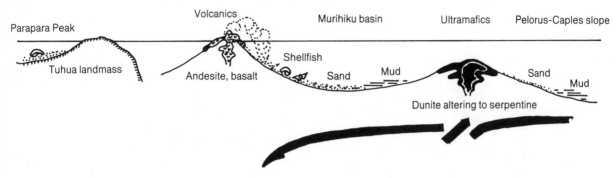

Fig. 86. Events during the Permian period.

The diagram above and similar pictures on later pages show possible events in the areas where the Murihiku rocks were formed. They are very generalised and are based roughly on the events shown by the rocks near Nelson City.

The Permian Period: 280 to 235 million years ago

Because a whole chain of volcanoes was erupting at the beginning of this period we believe there was plate movement, with oceanic crust moving down below the adjacent crust (Fig. 86).

Lava flows from these volcanoes are pretty rare; they seem to have produced mainly tuffs — hardened layers of ash such as we see today along the Desert Road. Some of the ash may have fallen on land, but most fell in shallow, coastal waters and was mixed with sand (Fig. 87). Some of the sand-tuff layers contain shell fossils that date the rocks as early Permian.

Southland

The biggest build-up of volcanic rocks made the Takitimu Mountains (Fig. 88), where the pile of tuff layers is 14 kilometres thick and gently tilting to the east. The volcanic rocks also extend northward to the Livingstone Mountains and southward through the Longwood Range to the northern part of Stewart Island. Both the Longwood Range and Bluff Peninsula are comprised partly of volcanics and partly of plutonic rocks of the same age.

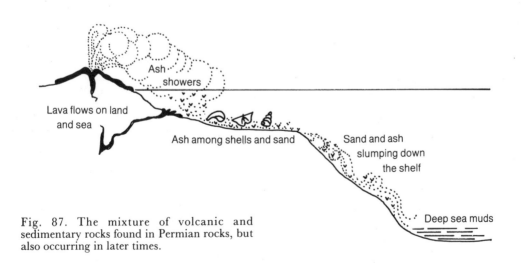

Fig. 87. The mixture of volcanic and sedimentary rocks found in Permian rocks, but also occurring in later times.

Fig. 88. The Takitimu Mountains near Redcliff, Waiau Valley.

On the east of Bluff Peninsula (Fig. 89), from Mokomoko Inlet to Ocean Beach, the sediments and ashes still show some of their original layering; a few patches of marble on the east of the inlet were formed from tiny bits of shell. Some larger shell fragments were found in the fine-grained volcanic sediments nearby, with one complete coral and one coiled shell. These fossils can be seen in the displays at the Southland Museum.

Most of the sedimentary rocks in this area were altered by heat when molten gabbro oozed into them in the middle Permian, about 245 million years ago. The gabbro cooled and solidified to a dark rock speckled with white feldspar, which now forms the Bluff Hill and can be seen along the coast (Fig. 90). It was quarried for the ornamental stone called Bluff Granite but which is in fact norite, a kind of gabbro containing hypersthene, an unusual variety of the iron-magnesium mineral pyroxene which can be distinguished from other pyroxenes under a petrological microscope (a microscope with various special accessories which distinguish minerals by the way their crystal structure alters light passing through them).

The border between the norite and the sedimentary rocks can be seen at Ocean Beach, on the shore opposite the freezing works, especially near the island with the larger boatshed. Here the blocky hypersthene crystals have been completely altered to a fibrous hornblende, a fairly common occurrence with norite. Here also the norite is layered and interleaved with sediments that have been completely recrystal-lised by the heat into a kind of gneiss, with banded coarse crystals of hornblende, feldspar, quartz and green epidote.

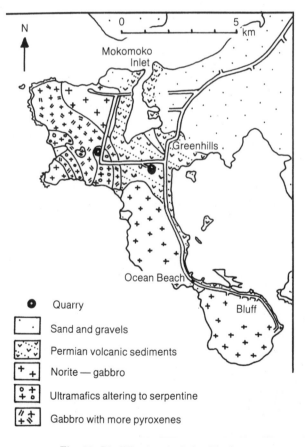

Fig. 89. Bluff Peninsula (*after Wood*).

Fig. 90. Bluff norite.

In the bay the tide washes over coarse, crystalline masses of glassy quartz and pink orthoclase feldspar made from the last liquids remaining in the magma. There are also seams of black and white speckled amphibolite (a hornblende-feldspar igneous rock) which run from shore to island. In the seams are found clusters of black hornblende with masses of dark-grey magnetite, a metallic iron oxide which is attracted to a magnet.

The banded gneiss on the beach grades into a less altered rock in the Greenhills area, where the sediments have recrystallised into "baked argillite", which is even-textured and very hard and provided excellent stone tools for the early Maori who worked this material. It has since been used for road metal, and large boulders are stockpiled in the quarry. Thin seams in the boulders show shiny flakes of dark-green chlorite, small cubes of iron pyrite and massive calcite with small crystals of quartz and feldspar and rarer sphene in holes in the rocks.

West of Greenhills the norite includes areas where the rock is made up largely of pyroxene crystals, and other areas where olivine is the main mineral. These form ultramafic rocks that have been partially altered to serpentine, which is being quarried for fertiliser. The working quarry is closed to collectors after the weekly blasting, but there is an old quarry further along the road.

The walls of the old quarry (Fig. 91) show vertical layers where seams of serpentine and minerals like fibrous, pale tremolite and flaky, dark antigorite have formed and are cut by later seams of more weathered minerals, including massive talc in the form of soapstone with a few tiny, octahedral chromite crystals. Crevices in the rock may contain white, radiating tufts of the calcium carbonate, aragonite, and similar tufts or rounded masses of the magnesium carbonate, hydromagnesite.

Fig. 91. Seams of tremolite and antigorite in the walls of the old serpentine quarry, Greenhills.

As the Permian volcanoes along the chain died down, the sediments in the nearby basins changed to limestones and sandstones containing various shellfish.

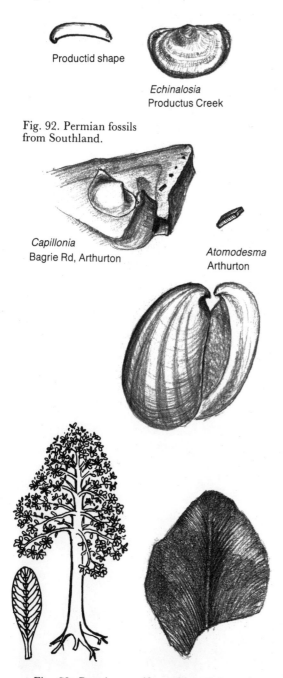

Fig. 92. Permian fossils from Southland.

Fig. 93. Permian seedfern, *Glossopteris,* and a leaf of *Glossopteris ampla* found at Productus Creek.

At Productus Creek, in a remote and isolated basin east of the Takitimus, a series of shell beds and limestones was discovered in the 1950s. It contains very good examples of an order of lampshells with incurving shells, the productids (Fig. 92), so the creek was named after them. Other layers up the valley side contain fossil corals and sponges and a grey limestone made up entirely of bits of shell, which is found also in the Eglinton Valley north of Te Anau and in the Maitai Valley in Nelson.

At Clinton and Arthurton, near Gore, there is a series of Permian outcrops in the farmland or on the edges of Bagrie Road where, with permission, you may grub among the cowpats and find pieces of brownish sandstone with small shell impressions which only a skilled paleontologist could decipher or love (Fig. 92).

In fact, the most characteristic Permian fossil is noticeable mostly by its absence. The shells of a bivalve, *Atomodesma* (Fig. 92), were composed of fibres built across the shell which caused it to break up into little blocks when the shellfish died. While the brick-shaped blocks sometimes make up the whole rock (the limestones at Mokomoko Inlet and Productus Creek), they are more often found scattered in sandstones. Since their lime dissolves quite readily in groundwater (these rocks have been underground for a long time), often all that is left is a brick-shaped hole. However, complete *Atomodesma* can be found. At Arthurton one outcrop contains shells showing stripes of their original colouration.

Some of this area must have been made up of sheltered water close to land, for among the shells and sand a few leaf fragments have been found, and at Productus Creek a whole leaf of the *Glossopteris* plant was discovered.

This is a special fossil. *Glossopteris* (Fig. 93) was a seed-fern tree which bore seeds like those of a pine but leaves like a bird's-nest fern. Fossil *Glossopteris* leaves have been found in all the southern continents, including Antarctica (Captain Scott hauled *Glossopteris* fossils on his last journey). Because these plants have not been found anywhere else and because they could not have reached South America from the north, their presence in all the southern continents is an important clue to the past unity of Gondwana.

Plate 1. Paleozoic and ultramafic rocks

A. Ruby in chrome mica, Hokitika (page 54).

B. Dunite with chromite and yellow rind (page 53).

C. Asbestos, Upper Takaka Valley (page 27).

D. Sapphire in tourmaline and chlorite, Golden Bay (page 36).

E. Grossular garnet, Maitai River, and Orepuki pebbles (page 54).

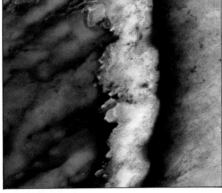

F. Nephrite jade from the West Coast (page 54).

G. Karamea Granite with pink and cream feldspar, black mica, grey quartz (page 42).

J. Blue kyanite, rusty quartz, black mica, Pariwhakaoho River (page 38).

K. Greenish sphene, purple-brown garnet, white apatite, Rameka Creek (page 43).

H. Red and green banded Permian siltstones, Maitai Valley (page 52).

I. Copper in serpentine, Champion Mine, Nelson (page 53).

L. Contact zone on the Rameka Track. Diorite on the left; pale, half-metre skarn at centre; grey marble right (page 43).

M. Five-metre band of dark Devonian limestone on the left, and sloping layers of sandy mudstones with some fossils to the right, on the banks of State Highway 7 by the Inangahua River near Reefton (page 34).

Plate 2. Torlesse Mesozoic greywacke and schist minerals.

A. Graded beds in schist, German Creek Bridge, Dansey Pass (page 136).

B. Garnet in gneiss, South Westland (page 78).

C. Magnetite in greenschist, Pivot Creek (page 78).

D. Scheelite and quartz in schist, Glenorchy (page 80).

E. Lustrous cubic galena; banded bluish stibnite; and soft, greasy, paler molybdenite.

F. Rocks at Akatore, stained black with manganese oxides (page 80).

G. Brassy chalcopyrite, yellow heavy gold and paler cubic iron pyrite.

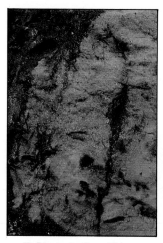

H. Rhodonite from Akatore (page 80).

I. Pillow lavas at Red Rocks, Wellington (page 72).

J. Radiolarian chert and manganese in folds, Red Rocks.

K. Red and green volcanic silts, Red Rocks.

Plate 3. Cretaceous volcanics and minerals

A. Petrified wood, Mt Somers (page 92).

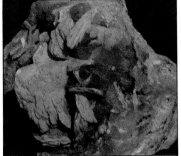

B. Green fluorite and cream barite, Thomson Hill (page 83).

C. Heulandite, a zeolite, Rangitata River (page 92).

D. Pillow lavas and zeolites, Ngahape (page 102).

E. Agate in andesite, Rakaia Gorge (page 91).

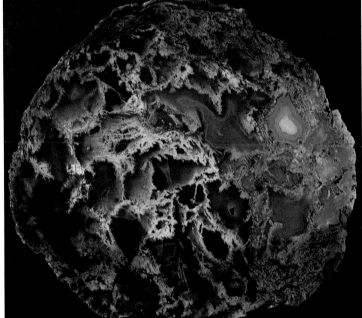

F. Moss agate, Clent Hills, Canterbury (page 91).

G. Green torbernite and yellow autunite, from the Hawks Crag uranium ore (page 82).

H. Ash and lava, Lookout Volcanics, Castle Creek, Awatere Valley (page 92).

I. Amethyst geode, Rakaia Gorge (page 91).

Plate 4. Rocks of the Cretaceous and Tertiary seaway, East Coast

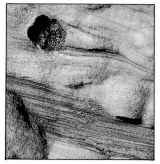

A. Pyrite nodule in bedded sandstone, Mangakuri Beach (page 101).

B. Detail of Waipawa chocolate shale (page 111).

C. Calcite in concretion, Moeraki Boulders (page 120).

D. Barite crystals in concretion, Mangakuri (page 102).

E. Cretaceous-Tertiary boundary in white limestone, Woodside Creek (page 108).

F. Cretaceous saurian bones in concretion, Amuri Bluff (page 104).

G. Chocolate and white mudstones, old bridge, Waipawa (page 111).

H. Petrified driftwood in conglomerate, Kaiwhata River (page 102).

I. Cretaceous sandstones and Paleocene limestone, Amuri Bluff (page 104).

Plate 5. Tertiary rocks formed during the incursion of the sea

A. Fossil sea urchin in greensand, Waihao Downs (page 125).

B. Calcite crystals, Browns Quarry, Southland (page 130).

C. Algal-crusted pebble, Mathesons Bay (page 146).

D. Serpentine and limestone quarry, east of Wairere Falls (page 141).

E. Oil shale outcrop, Nevis Valley (page 134).

G. Blue Spur Conglomerate, Gabriel's Gully (page 88).

F. Coal measures and grits overlain by greensand and mudstone, Fairfield (page 155).

H. Waipara Gorge showing the sequence from Eocene greensand to Miocene limestone (page 111).

Plate 6. South Island Tertiary volcanics

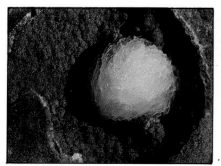

A. Calcite in basalt, Papanui Beach (detail) (page 157).

B. Opalite, Puddingstone Rock (page 157).

C. Aragonite, Sandfly Bay (page 157).

D. Crystals in basalt, Papanui Beach (page 157).

F. Columnar basalt, Waipiata (page 137).

E. White rhyolite under black basalt near Gebbies Pass, Banks Peninsula (page 157).

G. Pillow lavas, Boatmans Harbour (page 121).

H. Kakanui Breccia, north of Kakanui (page 122).

Plate 7. The latest sediments, formed during the final uplift and the Ice Age

A. Gold miners' workings, with gravels removed from the Miocene "Blue Bottom", West Coast (page 168).

B. The Wairau Fault, with offset outwash gravels, on State Highway 63 near Tophouse (page 194).

C. Brown weathered outwash gravels under younger, fresher layers, Rakaia Gorge (page 194).

D. Miocene sandstone beneath Pliocene limestone, Cape Turnagain (page 166).

E. Fox Glacier and the Southern Alps reflected in an ice kettle pond, showing the fault front (page 192).

F. Loess over basalt, Caroline Bay, Timaru (page 193).

G. Miocene-Pliocene mudstones and volcanic tuffs, Mangapoike River Valley (page 170).

Plate 8. North Island Volcanic rocks and minerals

A. Stibnite and quartz in sinter, Puhipuhi (page 201).

B. Olivine in andesite, Mahuia Falls, Tongariro National Park (page 215).

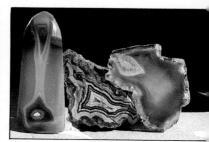

C. Carnelian from Te Mata, Coromandel (page 154).

D. Collecting rocks from a sinter (hot springs deposit) at Puhipuhi (page 201).

E. Siderite on quartz, Tui mine, Te Aroha (page 153).

F. Lead, copper, zinc ore from the Tui Mine (page 153).

G. Pink manganese calcite from Tararu Creek, Thames (page 154).

H. Azurite, Parakao, Northland (page 145).

I. Sulphur by the Champagne Pool, Waiotapu (page 217).

J. Basalt and rhyolite in the crater of Mt Tarawera (page 209).

Nelson

Here the volcanics are in the form of layers of ash, small bits of basalt and sandstones and mudstones with ash, all very much altered from their original appearance in the 260 million years since they were laid down. (See Fig. 94.) They can be seen in two small quarries on the main road (State Highway 6) at the north end of Nelson Haven, in the bed of the Maitai River

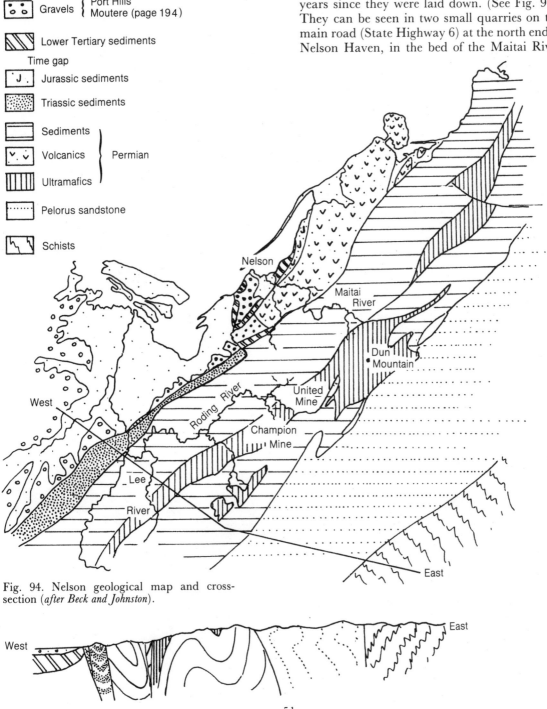

River gravels

Gravels { Port Hills
Moutere (page 194)

Lower Tertiary sediments

Time gap

Jurassic sediments

Triassic sediments

Sediments ⎱
Volcanics ⎰ Permian
Ultramafics ⎰

Pelorus sandstone

Schists

Nelson

Maitai River

Dun Mountain

United Mine

West

Roding River

Champion Mine

Lee

River

East

Fig. 94. Nelson geological map and cross-section (*after Beck and Johnston*).

West

East

between Cloustons and the Nile Street bridges and along Brook Street, from where they get their name, the Brook Street Volcanics. They sometimes contain fossilised shells, mostly fragments of *Atomodesma*.

In the Maitai Valley we can see the sediments that formed on the Permian sea floor after the volcanoes died down. In the lower valley limestones and sandstones contain a few shallow-water shellfish fossils, but further up the valley the rocks change to deeper-water sediments with thin layers of grey-green sandstones and siltstones, then very striking beds of alternating red and green siltstones (Plate 1H). They split cleanly along the layers and have been used often for garden walls and paths (Fig. 95).

These rocks show some sedimentary features — textures remaining in the rocks from the time they were deposited. Perhaps the most obvious is the ripple bedding visible well up in the Maitai Valley. The ripples in the rocks were made by water currents moving over the soft sea floor and were preserved when more sediment was deposited over them. Later the rock layers split apart along their rippled surfaces.

Permian ultramafics

At the end of the Maitai road is the geologically famous Dun Mountain (Fig. 96), an outcrop

Fig. 95. Permian sandstones from the banks of the Maitai Stream have been split and used for the bridge abutments.

Fig. 96. Dun Mountain from Maitai Valley.

made up almost entirely of the mineral olivine, from which the Austrian geologist Ferdinand von Hochstetter named the rock dunite in 1859. Because olivine has a high concentration of nickel, chrome and other metals, trees do not grow on Dun Mountain, and its bare, dun-brown mass is a landmark visible across Tasman Bay.

Dunite alters easily to serpentine, as we saw in the Cobb ultramafic rocks, where none of the original igneous rock remains. In Nelson and the other Permian ultramafic areas zones of serpentine surround the dunites or replace them completely. There are serpentine quarries for fertiliser in the Collins Valley and also in the Lee Valley where the magnesium carbonate, hydromagnesite (Fig. 97), can be found in tiny, white, radiating clusters, together with the spiky, white calcium carbonate, aragonite.

Dunite contains the heavy minerals chromite (Plate 1B) and platinum. These can sometimes be panned out of the streams or, especially in the case of chromite, they can be seen in the rock. Areas of Dun Mountain contain pyroxene as well as olivine (this type of rock is called peridotite.) Pyroxenes contain atoms of copper within the crystal, but when the pyroxenes in the ultramafics alter to serpentine the copper is forced out. It joins with any oxygen to make copper oxides (red cuprite or black tenorite) and with sulphur to make yellow chalcopyrite or brownish bornite, both metallic copper sulphides. Where oxygen or sulphur are missing blebs and crusts of pure copper are formed.

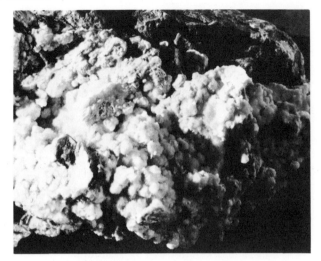

Fig. 97. Hydromagnesite from the Lee Valley serpentine quarry.

In the Nelson area huge masses of native copper have been found (Plate 1I), which led to mining in the 1880s when a smelter was built up the Aniseed Valley. The ruins can still be seen (Fig. 98) an hour's walk from the road end in the water reserve (two fords must be crossed). Tramways were built from the smelter to the mines in Champion Creek and United Creek. Rockhounds still make the extra hour's walk to the Champion Mine for the occasional specimen of native copper in serpentine and for thin crusts and microscopic crystals of the other copper minerals still to be found on the mine dumps.

Fig. 98. Smelter ruins, Aniseed Valley

As the ultamafics reacted with the host rock, limestone, seams of new minerals were formed, as in Rameka Creek. In the Dun Mountain area grossular garnet is found in translucent green masses that can be polished; with it are large pyroxene crystals, silky greenish-bronze diallage and darker enstatite. Rocks that are a mixture of grossular and diopside or diallage are known as rodingite, named for the Roding River, where specimens of these rocks can be found. They are also present in the Lee and Maitai Rivers.

Ultramafics appear further south, at Red Hill in the upper Wairau Valley and at Red Hills Range in north-west Otago south of Jacksons Bay, where there is a sizeable asbestos deposit. Serpentine is quarried at Mossburn in Southland and rodingite can be found in Cascade Creek in the Eglinton Valley on the way to Milford Sound. From there the grossular garnet makes its way through the lakes and down the Waiau, being tumbled into the rounded pebbles that can be found with a little effort on Orepuki Beach (Plate 1E).

The Permian ultramafics can be traced by the strong magnetic anomaly they produce — an alteration of the earth's magnetic field which has been mapped by geophysical surveys (Fig. 99). This anomaly runs from the serpentines of North Cape, down past the serpentine south of

Piopio (on State Highway 3 south-west of Te Kuiti), through the South Island and out to sea near Balclutha, showing where the rocks related to the Dun Mountain belt may now be buried. The surface ultramafics end in the Mavora Lakes area, south of Lake Wakatipu.

Certainly the ultramafics reappear along the West Coast mostly as serpentines with traces of asbestos included, but some of them are smeared and felted into the amphibole mineral nephrite, our greenstone or pounamu. From pods of nephrite within this serpentine, the West Coast rivers have sluiced boulders of jade and transported them to river beds or out to sea and along the coast, from the Hokitika River to the Taramakau and north to Barrytown Beach, where the diligent or the very lucky can find them.

As with most gemstones, jade is accompanied by a host of stones of very similar appearance, and most of the green stones in these rivers are serpentine or green-stained quartz. Nephrite jade can weather with a yellow rind, but it is too tough to be broken with a hammer. A few searchers find a waterworn piece with a slick, shiny, green surface.

One rather special rock of the greenstone belt is ruby rock or goodletite, a mixture of massive margarite mica stained rich green with chromium, fine, green tourmaline needles and tiny, dark-red rubies (Plate 1A). Unfortunately, this rock has been found only as rare boulders on the river plain and in dredge tailings near Hokitika. No one knows its source; it may be completely eroded away and no longer existing.

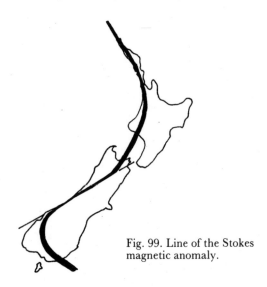

Fig. 99. Line of the Stokes magnetic anomaly.

To the east of the main Permian sediments there was a thick deposit of mudstones and sandstones building up in deeper water where very few shellfish lived, so the rocks contain almost no fossils. They occur in both islands, from South Auckland to South Otago, although their ages seem to vary. They are Triassic near Kawhia and at Tuapeka and Quoin Point on the Otago coast and Permian in the Humboldt Mountains and Caples River west of Lake Wakatipu where it is crossed by the Routeburn Track, and at Pelorus Bridge in the Marlborough Sounds. They are known as the Caples-Pelorus Group of rocks.

Fig. 100. Events during the Triassic and Jurassic Periods.

The Triassic and Jurassic Periods: 235 to 130 million years ago

During these periods the layers of sediment accumulated that now form the greater part of the Murihiku rocks (Fig. 100). The sandstones, siltstones and conglomerates of which they are made look almost the same from top to bottom. The layers are distinguished mainly by the changing groups of fossils within them. Although there was some renewed volcanic activity at this time, this is seen mainly as beds of tuff in the rock.

Southland

The rocks produced in the Triassic Period are divided into two series in New Zealand — Balfour and Gore — from the areas in Southland where these rocks are best seen (Fig. 101, p. 56). Each series is further divided into stages according to the evolving groups of fossils found within the rocks and has names from the type sections as follows:

Series	Stage
Balfour	Otapirian
	Warepan
	Otamitan
	Oretian
Gore	Kaihikuan
	Etalian
	Malakovian

The oldest Triassic rocks in the Southland Syncline are at the nose of the syncline, above Ohai in the Beaumont Hills. Here there are boulders of volcanic ash and debris containing small ammonite fossils. Ammonites are molluscs, members of the same group as the octopus and squid (cephalopods) but with an external shell. The only living member of the family is the chambered *Nautilus*, a member of the sub-class Nautiloidea, commonly called nautiloids. The Ammonoidea, a sub-class which includes ammonites, developed from this group, flourished and then died out while the nautiloids survived.

Nautiloids and ammonites were once very common and were spread across the seas of the world so that closely related species could be found in European and New Zealand waters. Their fossils are used to date the rocks in which they appear and they link New Zealand rocks with those formed in the same period in the rest of the world.

Nautiloids have chambered shells; the animal lives in the last chamber, forming a new chamber wall behind it as it outgrows the old one. The *Nautilus* of today has a simple, curved wall to its chamber, typical of all nautiloids (Fig. 102). They developed as far back as the Ordovician Period and were quite common in the Devonian, when their chamber walls generally had one sharp fold.

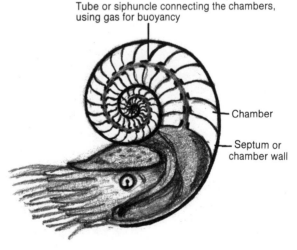

Fig. 102. Nautiloid in a chambered shell.

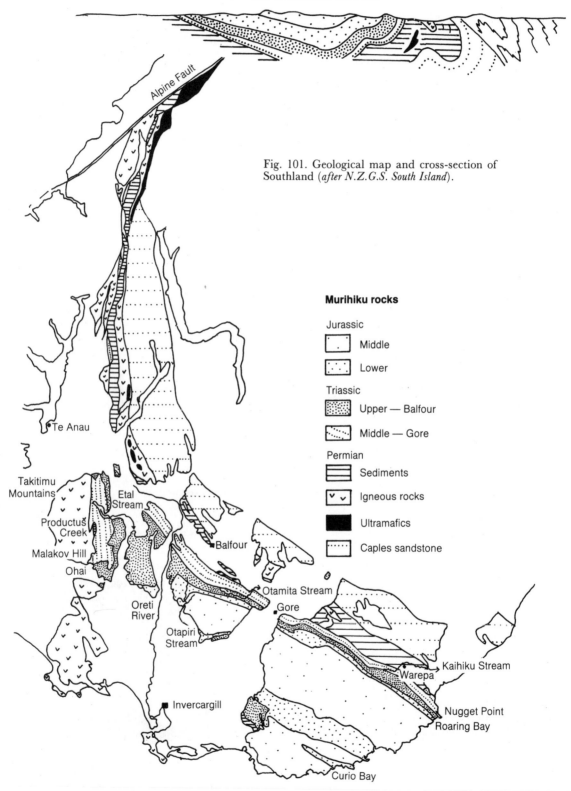

Fig. 101. Geological map and cross-section of Southland (*after N.Z.G.S. South Island*).

Murihiku rocks

Jurassic
- Middle
- Lower

Triassic
- Upper — Balfour
- Middle — Gore

Permian
- Sediments
- Igneous rocks
- Ultramafics
- Caples sandstone

But during the latter half of the Paleozoic, in the Permian, one group, the Ammonoidea, began to develop more complicated folds in the walls of their living chambers where they joined the outer shell (Fig. 103). These folds can be seen when the shell is removed or has dissolved away in the ground. At first the chamber walls were of a simple pattern, but by the Triassic the pattern resembled a delicate tracery of fern leaves. Many people seeing ammonites and their uncoiled relatives, baculites, for the first time think the tracery must be the print of fern fronds.

Every species had a different pattern, and the chambers distinguish the shells of cephalopods from the coiled shells of another group of molluscs — snails or gastropods.

The rocks of the lowest and oldest stages of the Triassic, the Malakovian and Etalian, are deep in the Beaumont Hills beyond Ohai on private property. Apart from ammonites the fossils there are not particularly attractive.

In the younger or higher levels of Triassic rock ammonites are not common. Brachiopods and bivalves are the more usual shellfish fossils found.

Good fossils of the Kaihikuan Stage (Fig. 104) can be found in boulders in the Etal Stream under the bridge on the good road from Mossburn to Nightcaps. Oretian fossils occur in the now-overgrown road cuttings near Dipton. Some of the best and most accessible Triassic fossils are on roadcuts near the Nugget Point

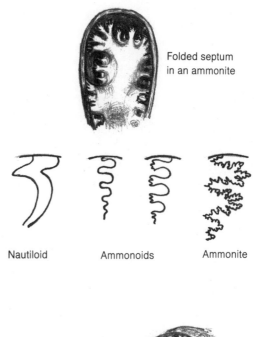

Folded septum
in an ammonite

Nautiloid   Ammonoids   Ammonite

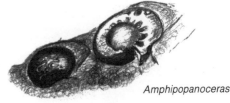

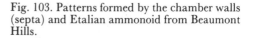

Amphipopanoceras

Fig. 103. Patterns formed by the chamber walls (septa) and Etalian ammonoid from Beaumont Hills.

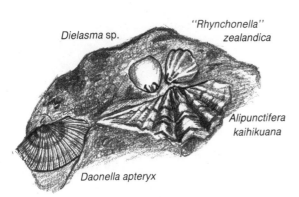

Dielasma sp.

*"Rhynchonella"
zealandica*

*Alipunctifera
kaihikuana*

Daonella apteryx

Halobia

Fig. 104. Left, Kaihikuan Stage fossils from Etal Stream. Right, Oretian Stage fossils from Dipton Cutting.

Fig. 105. Roadcut north of Nugget Point.

"Rhynchonella"

Alipunctifera kaihikuana
showing why some people call
brachiopods "stone butterflies".

Fig. 107. Kaihikuan Stage fossils from Nugget Point roadcut.

Fig. 106. Roaring Bay.

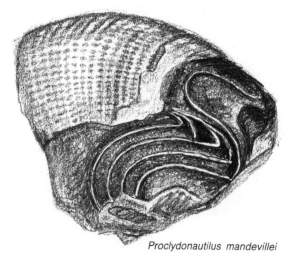

Proclydonautilus mandevillei

Clavigera bisulcata

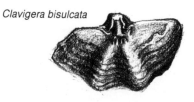

Fig. 108. Fossils from Roaring Bay. Left, Nautiloid showing net-like pattern on the surface of the shell; Otamitan Stage. Right, Brachiopod of the last Triassic stage, the Otapirian, from the rock wall at the end of the bay.

lighthouse on the east coast (Fig. 105) and along the sea cliffs just to the south in Roaring Bay (Fig. 106).

The top of the Balfour Series, the Kaihikuan Stage, can be seen at the cutting just as the road winds up to the track to the Nugget Point light; good specimens of the brachiopod *Alipunctifera kaihikuana*, as well as other brachiopods, have been picked up on the beach below (Fig. 107).

Going south along Roaring Bay the first fossils you find are likely to be Otamitan — brachiopods and a large type of nautiloid with a netted shell surface (Fig. 108). (Any perfect or unusual fossils should always be shown to geologists at a university or to the Geological Survey.) Some of the shell beds are full of one species, like the tightly packed layer of the bivalve *Manticula problematica*, which resembles a modern oyster bed (Fig. 109).

At the south end of Roaring Bay is a wall of Otapirian fossils which blocks the way to the Jurassic layers beyond the cliff (Fig. 108).

Inland, the Jurassic nose is at Ben Bolt, above Taylors Stream, where the rocks in the stream bed contain rare Jurassic ammonite fossils (Fig. 110).

Fig. 109. Tightly packed shells of the bivalve, *Manticula problematica*, of the Otamitan Stage.

Fig. 110. Taylors Stream below Ben Bolt, and the small Jurassic ammonite found in rocks in the stream bed.

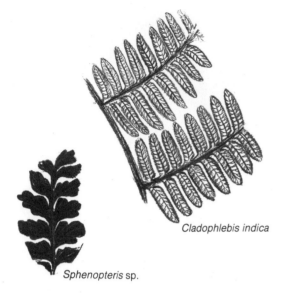

Cladophlebis indica

Sphenopteris sp.

Fig. 111. Jurassic plant fossils from Southland. Above, fern leaves. Below, petrified fern stems. Bottom, quartz crystals from petrified wood.

The most spectacular Jurassic fossils are at Curio Bay where a forest of trees lies fallen with the tree roots and stumps still in place on the shore platform. It is a fascinating locality, but what it offers is to be collected in photographic form only. Inland ditches and streams contain petrified wood and fern stems and imprints of leaves. The beach at Slope Point has rare punga nodules, which sometimes require a soak in bleach for the grain to show clearly. At Haldane Inlet, on the beach, there are pieces of wood, sometimes containing seams of crystal and agate. Punga is rare here (Fig. 111). Most of the wood is pale brown with a little black. Lapidaries search for fossilised wood in stream beds inland, in the Hokonui and Taringatura Hills. Plants in the Triassic and Jurassic were mainly conifers, seed-ferns and true ferns.

The presence of plant fossils in such quantity (members of rock and mineral clubs in Southland are specialists in petrified wood) show that large areas of land must have risen above the Jurassic seas. Perhaps these movements were the first stirrings of the great uplift to come?

Nelson

Here the Triassic and Jurassic rocks are found in tilted slivers between parallel faults, but some of the fossils of these periods are plentiful, and the slopes of Mount Heslington, at Richmond, were well combed by early paleontologists, including Hochstetter. *Monotis richmondiana* (Fig. 112) and other members of the genus are the type fossils defining the Warepan Stage and missing from the Roaring Bay sequence. Most of the Nelson fossils including *Monotis* are on private land on Mount Heslington and in 88 Stream.

Further north near Nelson City, at Marybank, plant fossils were identified as Jurassic as recently as 1980, showing that here also the Jurassic was a time when seas were shallowing and land areas spreading; although the Jurassic area preserved is merely a very thin strip between faults.

Monotis sp.

Fig. 112. *Monotis* shells may fill the rock or be present as a single cast.

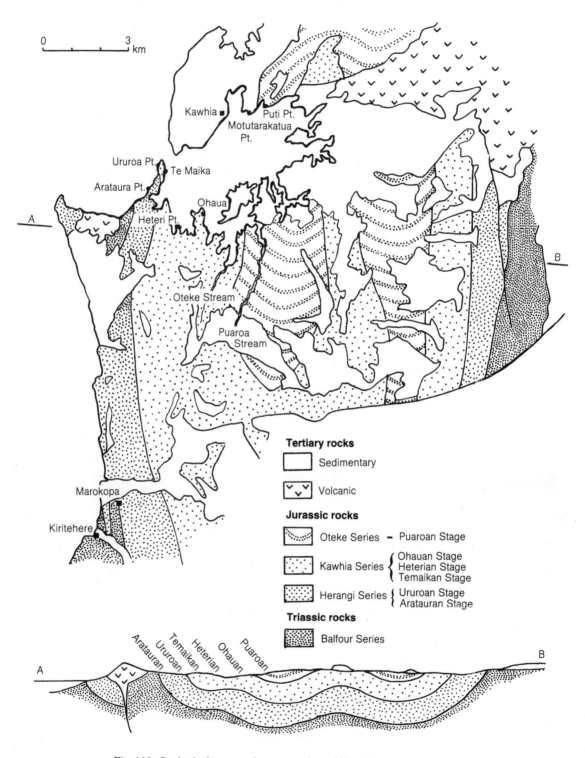

Fig. 113. Geological map and cross-section of Kawhia area (*after Kear*).

South Auckland

Ever since Hochstetter visited Kawhia in 1859 and found the first ammonites from New Zealand this has been our foremost Jurassic area (Fig. 113). Of course, the type sections are here — the Herangi Series is named for the mountains to the south and the Kawhia Series for the harbour rimmed with these rocks.

Although the Kawhia Syncline contains some of the same set of rocks as those of Southland, the Permian rocks are buried and those of the Triassic are out on the coast with access by sea or further south at Marokopa and Kiritehere, where *Monotis* shells of the Warepan Stage pack the rocks to the south of the beach.

The Jurassic rocks were mapped around Kawhia Harbour at low tide using boats. The rocks are progressively younger eastwards from the coast, and each layer containing distinctive fossils was named for a place on the shoreline. The series is reversed inland to the east on the other side of the syncline.

On land the oldest and most easily accessible rocks are those of the Temaikan Stage on the old road from Kinohaku to Taharoa, just north of the junction with the new Taharoa — Te Anga road. Further north along this road are Heterian fossils in the concretions, including the large clam *Inoceramus galoi* (Fig. 114). Many species of *Inoceramus* have proved most useful to geologists

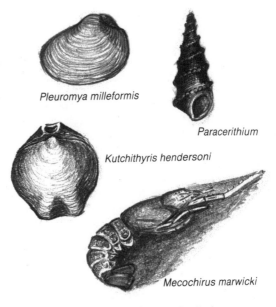

Pleuromya milleformis

Paracerithium

Kutchithyris hendersoni

Mecochirus marwicki

Fig. 115. Heterian Stage fossils from Whakapirau Road.

in separating the layers of the Jurassic and Cretaceous rocks in New Zealand. Here in Whakapirau Road many of the clams' shells have been replaced by shiny laths of the white mineral laumontite. Small shells in abundance are weathering out of the fracturing siltstones in the banks (Fig. 115).

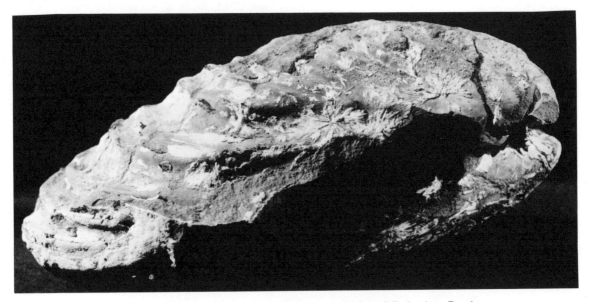

Fig. 114. Laumontite crystals on *Inoceramus galoi* from Whakapirau Road.

A few years ago an amateur collector recognised the edge of a giant ammonite in one of these roadcuts (Fig. 116). It is now on display in the Geological Survey in Lower Hutt and is possibly the largest Jurassic ammonite in the world and remarkably complete. The Waitomo District Council lent their quarry manager to blast it out of the solid, enclosing rock and geologists spent many months restoring it.

Curled up in the concretions nearby are fossilised fragments of small, crayfish-like creatures and rare complete specimens (Fig. 115). There are two types — one with claws and represented mostly by the head section, and the other *Mecochirus*, with a single extended pincer.

On the north side of the harbour are the Puti Point cliffs (Fig. 117), a little younger in age, where the shales fracture easily and contain many belemnites. These usually break, but with care they can be glued together.

Related to ammonites, belemnites were squid-like creatures (Fig. 118). Like the modern

Fig. 117. Puti Point, Kawhia Harbour.

Fig. 116. The giant ammonite, *Lytoceras*, from Whakapirau Road. *New Zealand Geological Survey Photograph by Diane Russell.*

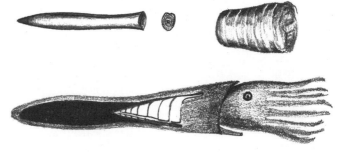

Fig. 118. Above, belemnite guard, chambered section and reconstruction of the squidlike animal. Left, belemnite weathering out of shale at Puti Point.

Fig. 119. Layer of volcanic ash in the mudstone at Puti Point.

Spirula, whose white ram's-horn shell is frequently washed up on beaches, they had in their rear end an internal structure made of layers of calcite. This guard is the bullet-like fossil found in some Jurassic and Cretaceous rocks. Some concretions at Puti Point contain also the chambered part in front of the guard, reminiscent of the chambers of the straight nautiloids and ammonites.

Concretions here also contain fossilised wood, which was probably carried out to sea as driftwood before being buried in the marine sediments. A few ammonites have been found by the lucky. The cliffs are banded by layers of white volcanic ash — tuffs that have been deformed and twisted by the weight of sediment on top of them (Fig. 119).

Further north, at Waikato Heads, the cliffs and roadcuts may yield Jurassic fossils. Just south of the heads at Cape Huriwai is a large deposit of fossilised wood and leaves, mostly fern, reached by a long walk over farmland. The Cape Huriwai cliffs contain fossilised tree trunks standing upright with spreading roots. On the beach below are fossilised portions of fern stems and slabs of sandstone bearing the imprint of leaves of ferns and conifers (Fig. 120). At low tide it is possible to collect leaf fossils beside the road on the rocks at Waimate Bay, about 2.2 kilometres to the east of the shop at Port Waikato.

Here also the former sea bed had become land by the end of the Jurassic Period. By this time the Murihiku sediments were complete.

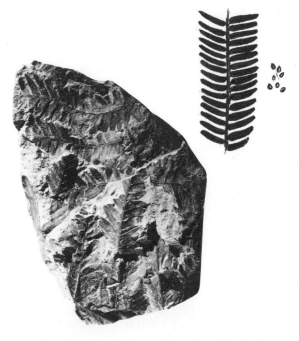

Fig. 120. Jurassic plants from Cape Huriwai, Waikato Heads. Photograph of fern, *Cladophlebis australis,* and drawing of a conifer and seeds.

THE TORLESSE ROCKS

Meanwhile, in another part of the sea floor, a different set of sediments had been building up without any trace of volcanic debris. It was made up mainly of layers of sand and mud which became the greywacke of the main ranges (Fig. 121). (Greywacke is a corrupted German word which may be used to describe hardened bands of sandstone and mudstone, regardless of age.)

The layers of mudstone accumulated slowly in deep water far from land. The coarser sandstones began as deposits much closer to land, but as they built up on the edge of the continental shelf they became unstable. Shocks such as earthquakes triggered off slumps. Big storms brought down from the land enormous amounts of sand and gravel which were swept far out to sea, perhaps down submarine canyons, taking with them some of the shelf material (Fig. 122). Thus, quantities of debris were carried far out on the deep ocean floor in massive flows called turbidity currents. The coarser material fell out of the current first and the finer muds took longer to settle down, producing a graded bed. As such masses of coarse material poured across the muddy sea floor they scoured out the mud and also ripped up pieces of mud and carried them along. Remains of such scouring and ripped-up mud fragments can be seen in those areas of the Torlesse rocks that show alternating beds of sand and muds (Fig. 123), such as the Welling-

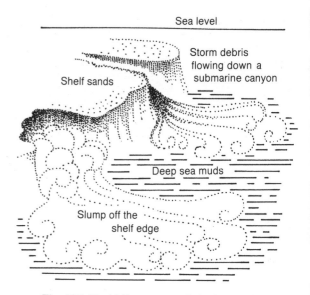

Fig. 122. Turbidity currents bringing sands to the deep sea floor.

ton greywackes at Titahi Bay and on the south coast of Wellington.

Greywacke sandstone is made up of grains of quartz, feldspar and iron-rich minerals such as hornblende, chlorite and biotite mica. The rock has many cracks. Air and water attack the feldspars, rotting them to form the clay mineral, kaolin. The iron in the other minerals rusts, and its rusty oxide, limonite, builds up along the cracks. The quartz grains and the clay wash

Fig. 121. Greywacke on Wellington coast.

Fig. 123. Textures in the Wellington greywacke. Left, graded beds with coarse sandstone bases to the right passing into darker mudstone. Right, ripped-up fragments of mudstone in sandstone.

away, leaving a network of limonite, which resembles a collection of brown boxes (Fig. 124). This occurs especially in coastal areas where salt aids the process.

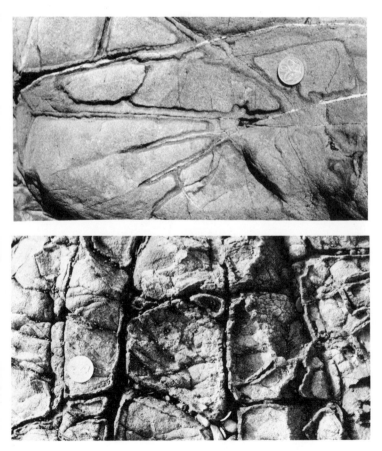

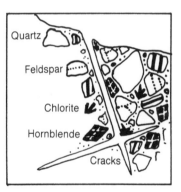

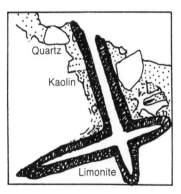

Fig. 124. Formation of a rusty boxwork weathering pattern in greywacke.

Torlesse fossils

Fossils in Torlesse rocks are not common, although more are being discovered. In the 3.9 million hectares of the South Island Torlesse there are fewer than 400 places where fossils have been found (Fig. 125) and many of these had only a single, river-carried boulder. In the North Island there are fewer than 100 Torlesse fossil localities. Many of the Torlesse fossils also lived in the Murihiku seas, which suggests that the two areas might not have been very far apart.

The oldest fossils are found on the south-east edge of the Torlesse at Kakahu near Geraldine, where the schists contain thick bands of marble. (The marble and younger limestones were once quarried and burnt in lime kilns like the one preserved by the roadside, Fig. 126.) The marble found here was once a limestone con-

taining conodonts (Fig 127), microscopic, tooth-like objects which appear after the marble is dissolved in acid. No one knows what animal produced the conodonts, but we can still use them to date the rocks in which they are found. Their shapes evolved over the ages, and by looking at the ages of fossils found with similar conodonts in other parts of the world we know the Kakahu conodonts are Carboniferous, the only fossils of this age found so far in New Zealand.

Permian fossils in the south are mostly

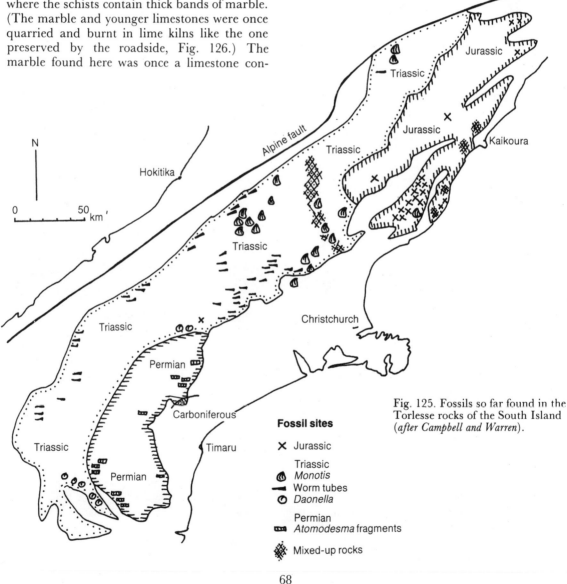

Fig. 125. Fossils so far found in the Torlesse rocks of the South Island (*after Campbell and Warren*).

Fossil sites

✗ Jurassic

Triassic
🐚 *Monotis*
▬ Worm tubes
◔ *Daonella*

Permian
▦ *Atomodesma* fragments

✿ Mixed-up rocks

Fig. 126. Kakahu lime kiln.
Photograph G. G. Thornton.

Fig. 128. Permian coral and fusilinids in marble, Marble Bay, and enlarged drawing of a fusilinid shell.

Fig. 129. Pale grey marble beside reddish-brown volcanic rock at Marble Bay, Northland.

Fig. 127. Carboniferous conodonts.

Atomodesma fragments, the same as those in the Murihiku rocks. However, one boulder near Benmore in the Waitaki Valley was found to contain fossil fusilinids, one-celled creatures which built shells rather like large grains of wheat. They were of the order Foraminifera (see page 108) and lived only in warm Permian waters.

Fusilinids have also been found in Northland, together with fossilised corals, crinoids and bryozoans — the animals of a coral reef (Fig. 128). The pale grey limestone formed by these reef creatures is found east of Whangaroa, Northland, where a track leads over the head-

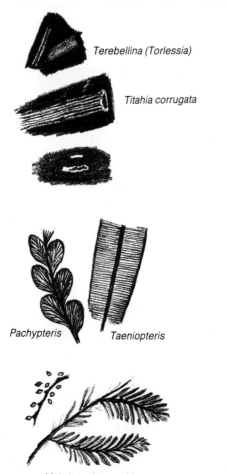

Terebellina (Torlessia)

Titahia corrugata

Pachypteris   Taeniopteris

Maitaia podocarpoides

Fig. 130. Torlesse fossils. Top, Triassic worm tubes. Bottom, Jurassic plants from the Clent Hills.

dissimilar rocks is discussed shortly, on page 73.

Triassic fossils in the Torlesse include the same shallow-water shellfish that thrived in Murihiku areas. However, the most widespread Triassic fossils in the Torlesse regions are not present in the Murihiku rocks. They are worm tubes (Fig. 130), generally found on the surface of black mudstone layers where worms lived on a muddy sea floor which may have been any depth from that of an estuary to the deepest ocean.

These tubes are of various kinds. The most widely found is *Terebellina* (commonly called *Torlessia* in New Zealand), a smooth tube which sometimes collapses and seems ribbed and is about 15 to 50 millimetres long. *Titahia* has a definitely ribbed surface and can be very much longer. At Titahi Bay, Wellington, other small, white, segmented tubes of cemented grains have been found in the sandy layers of the rocks.

Jurassic Torlesse fossils are mostly found in the north and east from the top of the North Island to North Canterbury. All are similar to the Kawhia fossils, and they include open-water animals, ammonites and belemnites as well as plant fossils, showing that some of the Torlesse also must have been lifted above the seas. Plant fossils from the Clent and Malvern Hills in Canterbury have been known about for some time (Fig. 130). More recently they have been found in the Aorangi Mountains in the Wairarapa.

Using all these fossils as time indicators, geologists have roughly divided the Torlesse rocks into areas of different ages.

Origins of the Torlesse rocks

A term often used in association with our Torlesse greywackes is geosyncline — a great, long trench in the ocean into which vast masses of sediment were dumped and which later rose to form new mountains. The New Zealand Geosyncline was assumed to have been such a trench, existing for some hundred million years, from the Carboniferous to the late Jurassic Periods. However, as more detailed work is done on the Torlesse, alternative views are being put forward.

There are more Torlesse rock types than the classic sandstone-mudstone sequences. There are areas of thick, black mudstones, and in some places massive sandstones with lenses of con-

land from Tauranga Bay to Marble Bay to the east. On the further end of this beach and also in neighbouring Oroua Bay are blocks of limestone, most of which is altered to marble, so the fossils on the weathered surfaces vary from almost recognisable finds to very dubious identifications made after patient searching. (Drawings usually show slightly interpreted versions of these fossils.)

The limestone blocks are enclosed in red and green volcanic rocks in patches along the shore and are also present as loose boulders, all between outcrops of greywacke sandstone (Fig. 129). The significance of this patchwork of

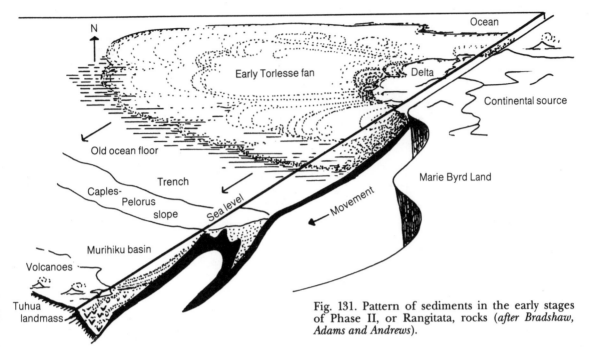

Fig. 131. Pattern of sediments in the early stages of Phase II, or Rangitata, rocks (*after Bradshaw, Adams and Andrews*).

glomerate are common. Many of these rock types could have come from relatively shallow water or from where a submarine fan was building out into deeper waters. Some of the fossils seem to have lived in shallower waters. In addition there are isolated patches of volcanics (red rocks) and limestones.

In the South Island two further factors have been considered in the search for a source for the Torlesse sediments: the quartz-rich nature of the sandstones and the absence of dark, volcanic material within them; and the location of coarser material in the south-east, where conglomerates and plant debris suggest a nearby land mass.

Given that the dark Murihiku volcanoes lay between the Gondwana continent to the west and the Torlesse seas, the granitic material that formed the Torlesse sandstones and mudstones could not have come from the Gondwana continent in the west. It could not have jumped the volcanic barrier and arrived in such a pure state. The granitic debris must have come off

another part of the Gondwana area thought by some geologists to be Marie Byrd Land in Antarctica. Fig. 131 illustrates a recent impression of the whole New Zealand scene when the Rangitata rocks of the second phase of our geological history were being formed.

In the north of the country the influence of granitic debris in the Torlesse rocks is lessened and the greywackes north of the Waikato do contain volcanic material. We have already looked at the patches of limestone mixed with red volcanic material in Marble Bay; such areas are quite common in the Torlesse, although the limestones do not always contain tropical corals or other warm-water fossils. Red and green rocks show up on the western side of the Rimutaka Hill, on State Highway 2 from Wellington to the Wairarapa, and at Red Rocks on the south coast of Wellington, west of Owhiro Bay. Oxidised iron in the volcanics makes these rocks red. When less oxygen is available iron minerals are greenish and a mixture of both colours is often seen.

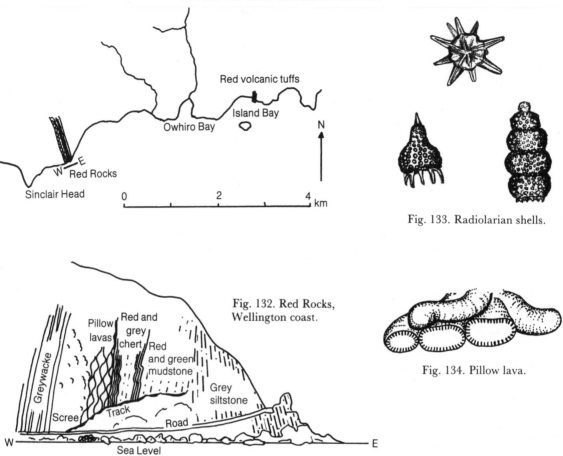

Fig. 132. Red Rocks, Wellington coast.

Fig. 133. Radiolarian shells.

Fig. 134. Pillow lava.

At Red Rocks (Fig. 132) there are very distinct layers. The grey siltstone shows cleavage and contains patches of lime. The mudstone is coloured from iron in hot volcanic brines. These also contained silica that encouraged the growth of vast quantities of Radiolaria (Fig. 133) — minute, one-celled creatures whose skeletons contained silica — and, from their remains, beds of the rock chert built up in the tuffs. At one stage lava erupted from fissures on the sea floor. As it emerged its outer surface was chilled by the water to form a coherent but still plastic envelope which filled with molten lava to form a pillow-like structure. The repeated occurrence of this produced a stacked layer of these structures, termed pillow lava (Fig. 134).

In most red rock outcrops like this there are faults between the red rocks and the surrounding greywacke; the volcanics do not cut across the layers of sandstone, nor do the sandstones contain volcanic materials.

Fig. 135. Manganese oxide from North Auckland.

Linked with the red rocks in the Torlesse are deposits of metals — copper at Maharahara near Woodville and at Kawau Island, and manganese, particularly in Northland and the Hunuas (Fig. 135). In fact, it was the manganese that first brought the miners to Kawau Island; the copper was discovered and worked later.

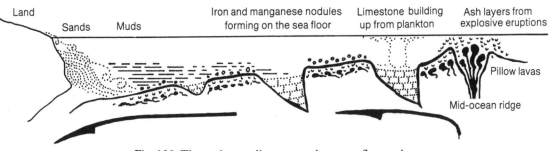

Land

Sands   Muds   Iron and manganese nodules   Limestone building   Ash layers from
                 forming on the sea floor   up from plankton   explosive eruptions

Pillow lavas

Mid-ocean ridge

Fig. 136. The various sediments on the ocean floor today.

On the ocean floor today mid-ocean ridges are fissured, and pillow lavas erupt and metallic brines spread from these gaps into the surrounding waters, leaving sulphide deposits on the ocean floor. As the crust moves away from a ridge different sediments accumulate on the ocean floor, forming layers composed of materials that were created in quite different environments and times (Fig. 136).

It is likely that the northern sediments and a good proportion of the Torlesse rocks could have formed in the same way — normal ocean deposits being scraped off the ocean floor on to the toe of an overriding plate where they would meet up with sediments coming off the adjacent land. The slow-moving oceanic crust could have carried together pillow lavas, ashes and cherts from the oceanic ridge, manganese in nodules such as those built up on the ocean floor at the Chatham Rise (Fig. 137), limestones from high spots, and folded them in with sandstones and mudstones from the continental shelf.

The concept of a single deep trench, the New Zealand Geosyncline, may therefore be replaced by the idea of a far-flung, shallow carpet of sediment, which was folded together and rumpled up (Fig. 138).

This movement may have taken place in several stages. We accept that a great uplift occurred at the end of the second phase of our geological history, but perhaps this was merely the grand finale after several lesser events?

The evidence for an earlier pile-up comes from the South Island. Between the older and younger Torlesse rocks there is a mixed-up zone of rocks of both ages, and there is a time gap shown in the fossils where no sediment of the early Jurassic Period has been found. Maybe the first great collison occurred at the end of the

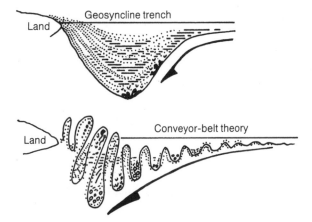

Geosyncline trench

Land

Conveyor-belt theory

Land

Fig. 138. Two ways to make the Torlesse sediments.

Fig. 137. Manganese nodules on the ocean floor at 5720 m depth. (*Photograph courtesy of the Oceanographic Institute, D.S.I.R.*)

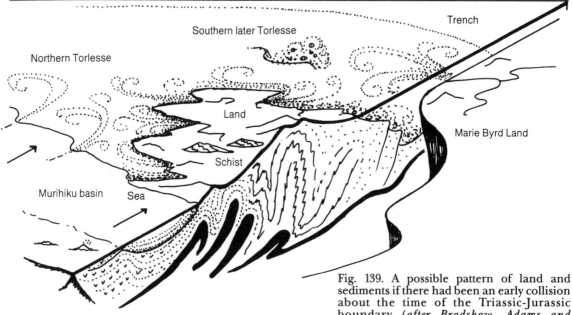

Fig. 139. A possible pattern of land and sediments if there had been an early collision about the time of the Triassic-Jurassic boundary (*after Bradshaw, Adams and Andrews*).

Triassic, producing patches of uplifted land where later Jurassic plants could have grown and shallow water fossils could have colonised new shores? In the remaining ocean areas the rest of the Torlesse could have accumulated and chunks of the older land could have been swept down (Fig. 139).

Probably, in the first collision outlined above, the older Torlesse layers rumpled up and pushed against the edges of the Caples-Murihiku rocks, welding the two together and shoving down an enormous depth of sediments, which were altered completely to form the last main group of the Phase-II rocks.

THE HAAST SCHIST — REGIONAL METAMORPHISM

Between the Torlesse and the Murihiku in the South Island is a great belt of altered rock, mainly schist (Fig. 140). We have already seen how rocks may be altered by contact with other hot, igneous rocks and some of the results of uplift and pressure in the formation of marble and gneiss in our oldest areas. But the schists of the South Island cover many hundreds of hectares. A regional alteration of this magnitude must have been the result of deep burial under a pile of sediments.

At the south and east edges of the belt the rocks pass into normal sediments, sandstones and mudstones of the Torlesse or Caples Groups. But when geologists first explored and described these rocks they saw increasing alteration as they moved inwards from the edges of the belt, marked by a series of differences first described in Scotland and then refined in Otago. Now the changes have been put into a definite series, so anyone in the field can sort out these rocks by their degree of alteration without needing special laboratory tests.

We would need microscopes to see the first changes that have occurred. They are in the form of tiny minerals which began to grow between grains of the original sediments without much alteration to their texture. Rocks exhibiting this first degree of alteration are not usually regarded as metamorphic at all, since most of the older greywackes contain these new minerals. They include the green calcium-aluminium silicates, pumpellyite and paler prehnite, and some zeolites which form white veins in the rocks.

The most obvious sign of metamorphism is the cleavage developing in the rock. As the rocks were pressed down with the weight of the overlaying rock, the new minerals began to grow on the surfaces most affected by the pressure,

Fig. 140. The Haast Schists (*after Suggate, Stevens and Te Punga*).

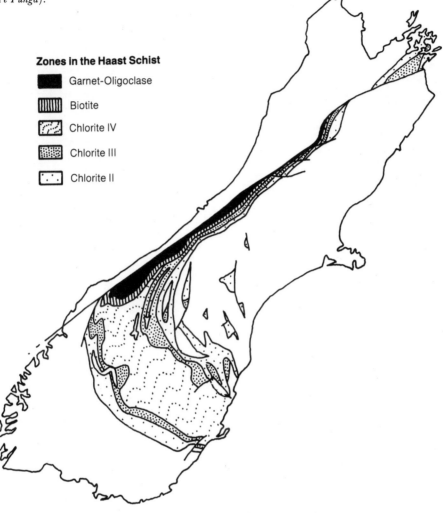

Zones in the Haast Schist

- Garnet-Oligoclase
- Biotite
- Chlorite IV
- Chlorite III
- Chlorite II

looking as though they were squeezed flat. They provide an alignment in the rock (Fig. 141) and make it split easily. Micas, like silvery muscovite and dark-green chlorite, are the commonest

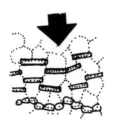

Fig. 141. Alignment of crystals growing in a metamorphic rock in response to pressure.

flaky minerals and two of the first to grow in pressurised rock.

With increasing pressure the minerals grow larger and begin to separate into light and dark layers and the rock becomes recognisable as a schist. Most of Otago is composed of schist in the chlorite grade, zoned according to whether it has cleavage but no distinct layering (Chlorite II); distinct layering with aligned mica plates (Chlorite III); or foliation, separation into marked white layers (mainly of quartz and feldspar) and dark layers of mica (Chlorite IV), as seen on the shores of Lakes Wakatipu and Wanaka. See Fig. 142.

Fig. 142. Degrees of alteration in schist. Left, blocky cleavage but no layering in Chlorite II schist at Fairlight, south of Kingston. Right, well-defined layering in Chlorite IV schist near Queenstown. Bottom, the sequence from Chlorite II at left to garnet schist on the right.

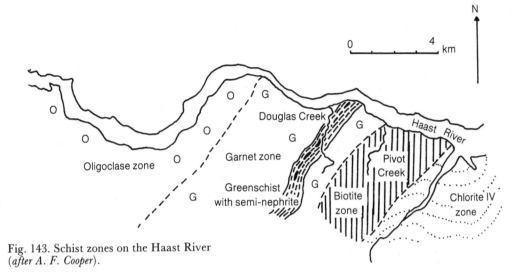

Fig. 143. Schist zones on the Haast River
(*after A. F. Cooper*).

The rocks are much more strongly altered along the banks of the Haast River where the river turns west to the sea (Fig. 143), and the sequence here provides the name for the whole group — Haast Schist. Black mica, or biotite, first appears in these rocks 1.5 kilometres west of the turning corner. It continues to be part of the rock lying westwards until it reaches the place where garnets also appear, with a more calcic feldspar, oligoclase, coming in to the west. However, since oligoclase cannot be distinguished from the eastwards feldspar, albite, except under a microscope, the oligoclase boundary is sometimes aligned with the garnet boundary for ease in mapping.

These changes in the rocks were first seen and mapped before theories could be devised to explain them. Since localised heat and pressure could not affect a whole region it has now been accepted that the entire area was once buried deep under piles of other rocks; deep burial results in high pressure and heating of the rock.

Overseas, miners have always found that the deeper inside the earth they went the hotter the mine became. The deepest drill-hole made, down to some 8 kilometres, rose in temperature some 20°-30°C per kilometre. This means that rocks buried under 20-30 kilometres of younger sediments could reach temperatures from 400° to 700°C. At these temperatures quartz and feldspars melt, forming granites in the deeper

part of the crust. But long before the rocks melt other changes take place.

As temperatures and pressures increase, various minerals become unstable and their atoms move to regroup and grow into new minerals. This occurs within a solid rock, with each mineral or group of minerals breaking down or reacting in turn.

These reactions have been studied in laboratories. Sediments like the typical greywacke sandstones of the Southern Alps have been enclosed in solid containers and pressure-cooked to different temperatures before being examined to see what new minerals have formed. After repeated experiments geochemists have discovered the conditions that will produce or destroy all the minerals and textures so far seen in our schists.

As a caterpillar metamorphoses to a butterfly, dull, grainy sandstone changes to a shimmering, garnet-studded schist. From the minerals seen in the rocks we know the temperatures they must have reached and can estimate the depth of their burial (Fig. 144).

Of course, the actual minerals produced depend upon the chemistry of the original rocks. A rock rich in iron, magnesium and calcium (such as basalt) will make a greenschist containing dark-green chlorite, bright lime-green epidote and lathlike black hornblende at about the same metamorphic grade as garnet. The extra

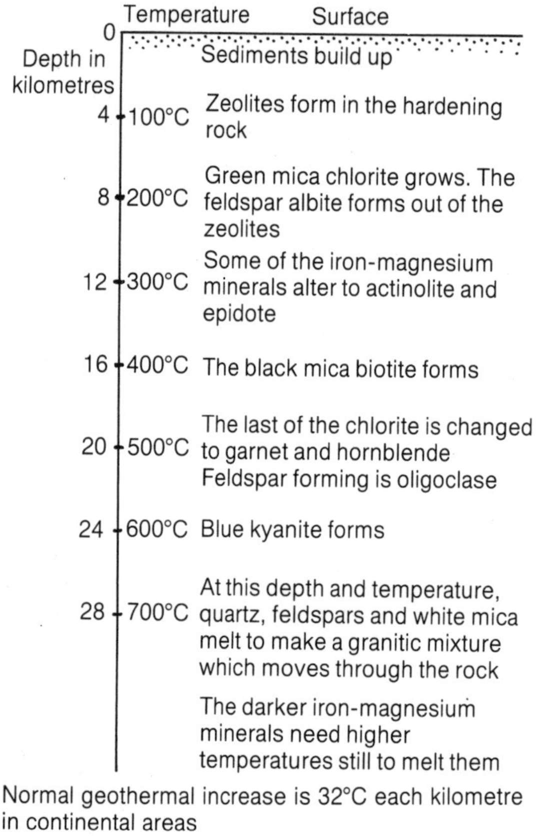

Temperature   Surface

Depth in kilometres

Depth	Temp	
0		Sediments build up
4	100°C	Zeolites form in the hardening rock
8	200°C	Green mica chlorite grows. The feldspar albite forms out of the zeolites
12	300°C	Some of the iron-magnesium minerals alter to actinolite and epidote
16	400°C	The black mica biotite forms
20	500°C	The last of the chlorite is changed to garnet and hornblende. Feldspar forming is oligoclase
24	600°C	Blue kyanite forms
28	700°C	At this depth and temperature, quartz, feldspars and white mica melt to make a granitic mixture which moves through the rock
		The darker iron-magnesium minerals need higher temperatures still to melt them

Normal geothermal increase is 32°C each kilometre in continental areas

Fig. 144. Temperatures and depths at which the various minerals form.

iron often forms octahedral crystals of the iron oxide, magnetite (Fig. 145). In the greenschist bands in the Haast series there are also serpentines, which have almost become nephrite jade. Some of the schists also contain dark-blue laths of the amphibole riebeckite, a sodium-iron silicate.

Further north, in South Westland, at Hunt Beach and Jacobs River, boulders of schist may contain blue kyanite and the chrome-green mica, fuchsite, but they are rather rare. Garnets are more common (Plate 2B).

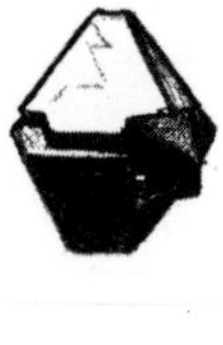

Fig. 145. Magnetite crystal.

Travellers from the North Island first see these schists in Marlborough as they drive from Picton to Blenheim through roadcuts in layered rock which shows a definite cleavage but contains very little visible mica. It is Chlorite II grade. It is a useful building rock, and at Langleydale (Fig. 146) the Maori built from it a rampart round their outlook over the trading route across the Wairau. Here the schist is Chlorite III, containing visible mica.

In Central Otago the early settlers found the easily split schist a substitute for the timber which was so conspicuously absent. They made fenceposts, buildings and walls from this rock (Fig. 147). Today it is used as a decorative building stone, especially the pink manganese schist from beyond Wanaka and the more common greenschist which is scattered throughout the region (Plate 2C).

Along the Southern Alps there is a narrow belt where the grades of alteration in the schists can be very clearly seen. Where the Fox and Franz Josef Glaciers have cut their way down from the alps, the different zones are visible in the walls of rock. They range from high-grade garnet-oligoclase rock right back to the almost unaltered sandstones of the high peaks. We can also see the way in which the isograds (boundaries of rocks of equal metamorphic grades) cut across the folded rocks (Fig. 148), suggesting that the rocks were folded as the overlying rocks were piled up; the folding suggests the burial was part of a collision pile-up, not just a normal accumulation of sediment.

Given that these rocks were once buried to great depths, they must have come to the surface by being pushed up along a break in the crust, namely the Alpine Fault (Fig. 149), the present boundary between the Pacific and Indo-Australian Plates.

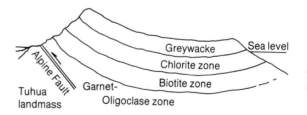

Fig. 149. Very simplified section across the lower half of the South Island.

Fig. 146. Maori rampart built in schist at Langleydale, Wairau Valley.

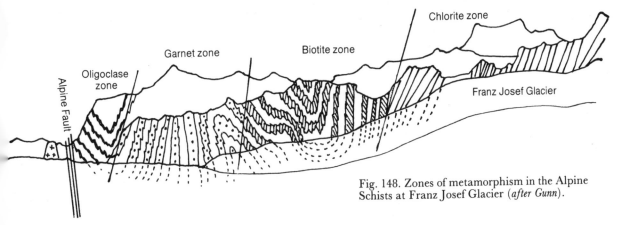

Fig. 147. Mitchell's cottage, Fruitlands.

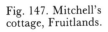

Fig. 148. Zones of metamorphism in the Alpine Schists at Franz Josef Glacier (*after Gunn*).

Fig. 151. Grey metallic blades of stibnite, antimony sulphide.

Fig. 150. Round Hill Mine, Macraes Flat.

As the schists developed, scattered metallic elements in the sediments came together to form lodes or ore-bodies. The most notable gave us the goldfields of Central Otago as well as lesser bodies in Marlborough. At Bendigo, Carricktown, Oturehua and many other places veins of quartz containing golden grains were discovered and mined. The tunnel illustrated in Fig. 150 at the Round Hill mines near Macraes Flat followed a lode in the schist near the top of the cut.

Over the ages flakes of gold were eroded out of lodes now vanished and were concentrated in rivers and old gravels from Central Otago to the sea, where they were panned, sluiced and dredged. Today some of the sites and relics are preserved for the public in the Central Otago Goldfields Park.

Not only gold but also other metals were concentrated. At Glenorchy and Macraes Flat the cream-coloured heavy tungsten ore, scheelite (Plate 2D), was found and is still being mined on a small scale. (Scheelite is a mineral, calcium tungstate). At Golden Point, near Macraes Flat, the stamper battery is being operated for gold and scheelite as part of the Goldfields Park; mine dumps stud the hillside behind it, and quartz crystals and occasional pieces of scheelite ore may still be found. Stibnite (Fig. 151), the grey, soft, bladed ore of antimony, was mined at Mount Stoker near Middlemarch and at Alexandra, on the south bank of the Clutha River just over the main bridge. Houses have been built there, so collectors would have to search carefully by the roadside for the yellow-stained rocks with small, grey, shiny blades. Stibnite was also mined at Resolution Bay and Endeavour Inlet in the Marlborough Sound, and scheelite at Wakamarina.

At Moke Creek, north of Twelve Mile Bay on Lake Wakatipu, there are thick seams of copper and iron pyrite paralleling the schists, but the deposits have been too small to be worked for any length of time.

The purple schists near Wanaka are a product of the manganese mineral piedmontite. Another manganese silicate, rhodonite, is found in massive form and makes an attractive gemstone, bright pink with patterns of black manganese oxide. Any really black boulder on the gold dredge tailings at Lowburn, Cromwell and Alexandra may be found to have a pink core.

The rocks described above are the altered equivalents of the metal deposits found in the Torlesse, such as the Woodville copper or the manganese rhodonite present on the beach between Akatore and Quoin Point, south of Dunedin (Plate 2F, 2H).

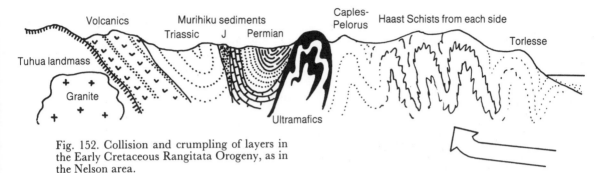

Fig. 152. Collision and crumpling of layers in the Early Cretaceous Rangitata Orogeny, as in the Nelson area.

THE RANGITATA OROGENY — GETTING IT ALL TOGETHER

The collision of the Murihiku and the Torlesse and their welding together as the sediments folded and altered to schist may have taken place in the late Triassic/early Jurassic Periods, but the final uplift and ending of all the Phase II sedimentary build-up certainly occurred in the early Cretaceous, some 130 million years ago. All the Phase II layers were compressed and lifted up and the land itself rumpled into large folds (Fig. 152).

Some of the evidence for the great uplift includes breccias (pronounced "bretchyas") at Hawks Crag in the Buller Gorge and at Kyeburn and Henley in Otago. (Breccia is a rock made up of rough pieces of rock cemented together [Fig. 153]; we can see broken pieces of this rock today in mountain screes on steep slopes.) The breccias at Kyeburn and Henley rest on eroded schist surfaces, which further suggests that the schists were formed before the great upheaval.

Fig. 153. Angular breccia and rounded conglomerate.

In the Buller Gorge the Hawks Crag Breccia shows that the land was steep and new; the scree deposit backs up to a hillside of older rocks.

Breccia appears also at the mouth of the Fox River (Fig. 154) by the main road bridge.

In the Buller a river plain seems to have built up of sandstones and shales with plant impressions, such as that which can be seen 200 metres upriver from Berlins Hotel in the lower Buller Gorge, about 10 kilometres west of Inangahua.

Fig. 154. Hawks Crag Breccia at the Fox River.

The plant impressions are mainly of fragments of horsetails, *Equisetites* (Fig. 155), a kind of rush which still grows overseas and has been reintroduced to New Zealand. Pollens found in these beds show they are about 95 million years old.

Fig. 155. Horsetail plant, rather like a pine seedling.

In 1955 uranium was found in the Hawks Crag Breccia, mainly as very fine, black grains of the uranium silicate coffinite, but in a few places collectors have found flakes of yellow autunite and green torbernite, other uranium minerals (Plate 3G). At present the deposits are not economic to mine.

During the time the land was being uplifted there was a general break in the series of marine sediments, although in Marlborough and the East Coast of the North Island the break was very short and occurs at different levels in different places. However, it does seem that nowhere in New Zealand is there a continuous set of beds formed from the Jurassic to the Middle Cretaceous, and no very early Cretaceous fossils have been found.

The folding of the land formed basins, which were covered by the sea and filled with sediment, and areas of high hills and plains. In Southland and Kawhia the folds are gentle, as illustrated on pages 56 and 62, but in Nelson the rocks were tightly squeezed until the sides of the folds broke to form faults. Some of the time periods are represented by only tiny slivers of rock from the east flank of the syncline; the west and more gently dipping flank is buried under the Moutere Gravels.

As in the Tuhua uplift, the Rangitata Orogeny was accompanied by the melting of granitic materials deep in the crust and their movement into the overlying rocks.

Most of the Cretaceous granites are in the easternmost belt, running down from Separation Point in Abel Tasman National Park to the east of Fiordland. They contain fewer pink feldspars than the older granites and seem to break down sooner. At Kaiteriteri Beach in Tasman Bay the golden sands come from the weathered granite, and out on the reefs the granites also contain small orbs (rounded balls of distinct layers of quartz, feldspar and mica, making the same orbicular granite as that up the Glenroy Valley). Feldspars in the granite break down to form clay minerals, giving Motueka the fine kaolin clay used by potters.

The Cretaceous granites also formed many small plutons or intrusions into the older granites, but only with very careful work may the Karamea and Cretaceous granites be distinguished. It is possible that the later granites introduced more minerals — we saw copper in the Pariwhakaoho River from a Cretaceous granite. In Canaan Valley on Takaka Hill there are scheelite (the ore of tungsten), and copper in the marble-granite contact zones as well as tourmaline and the red titanium mineral, rutile, and clusters of fine quartz crystals.

Mount Radiant, above the Little Wanganui River in North-west Nelson, contains a sparse amount of gold and silver with more copper and molybdenum, and down near Reefton recent exploration has uncovered tin and tungsten as well as gold.

High on Thomson Hill (Fig. 156), between the Baton and Wangapeka Rivers, there is a deposit of green fluorite, creamy barite and quartz crystals on a contact zone between marble, granite and older diorite (Plate 3B). The area has been well picked over considering the climb involved to get there.

The radiometric ages of the granites seem to fall within the 110-100 million years range, which gives us an indication of the duration of the later part of the Rangitata Orogeny.

Fig. 156. Above, crested barite, clustered fluorite and quartz crystals, Thomson Hill. Right, collecting at Thomson Hill. Below, quartz crystals, Canaan Valley.

CHAPTER FOUR

Phase III:
A place of our own

By the middle of the Cretaceous Period, some 100 million years ago, there was a new landmass on the edge of the Gondwana continent — a folded and mountainous land. The third cycle of erosion and the dumping of enormous quantities of sediment on lower areas began, but this time much of the sediment formed a blanket on top of the older land and only small areas of new land were built out on the eastern margin.

We sometimes need to remind ourselves that the Cretaceous Period lasted half of all the time that has elapsed since its beginning some 130

million years ago (Fig. 157). In the 65 million years of the Cenozoic Era, after the Cretaceous ended, so much happened in the development of our life and land that we have been able to use the changes to divide the Cenozoic into small, distinct epochs and we tend to forget just how long a span of time is covered by the word Cretaceous.

We will look at the Cretaceous rocks under three broad headings — the sediments built up on dry land, the volcanics erupted on the edge of the land, and the sediments deposited in the seas to the east.

The first part of Cretaceous time is not recorded in our rocks because it was the time of the great uplift. High lands leave no record of their life span except in the sediments they give to other areas. Millennia must pass before sediments begin to build up in new troughs and before uplands wear down to form lowlands which are covered with plant and river debris, making rocks which record the passage of time.

However, the last 40 million years of the Cretaceous *is* represented in our rocks and great changes have taken place during this time. Much of the new land has worn away, volcanoes have erupted and become extinct and the piece of the continental crust that forms New Zealand

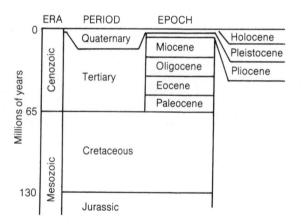

Fig. 157. Time divisions of the last 130 million years drawn to scale.

Fig. 159. The Cretaceous land.

has split from Australia to go it alone across the Pacific.

The actual shape of our land mass at that time is still uncertain. It would not have been the present shape of New Zealand because of the movement that has since occurred along the Alpine Fault, tearing the Nelson-Southland rocks apart. Because the fault system is still moving, many geologists believe the whole displacement of these rocks took place in the last 10 million years or less, so in their map of Cretaceous New Zealand the Nelson and Southland rocks are together. Other geologists believe most of the displacement of the Nelson-Southland rocks occurred in the Rangitata Orogeny, and their pattern of the Cretaceous basement and land is close to the present structure of New Zealand, with varying shorelines.

In the South Island the Alpine Fault is a clear feature. In the North Island the position of the break has been put either west of the island, west of the Kaimanawas, or in the ocean to the east of the island. Possibly there are three fractures and movement may have taken place on all three at different times. Some scientists would even have New Zealand split and turned inside out as they remove the recently created sea floor and try to reconstruct the oceans of the past (Fig. 158).

So far no single map meets with universal approval, which is why most of the diagrams in this book are in cross-section form and therefore do not depend upon map outlines.

SEDIMENTS ON THE CRETACEOUS LAND

Whatever the shape of the land, rivers ran from central mountains down to the sea. Plants grew in a dense bush on the flood plains by coastal swamps and lowland lakes. Slowly dead and decaying plant material formed thick deposits of peat in basins across the land (Fig. 159).

The layers of plant debris that began to build up in the Cretaceous made our oldest coal deposits which were mined in Nelson, Westland, Southland and Otago (Fig. 160).

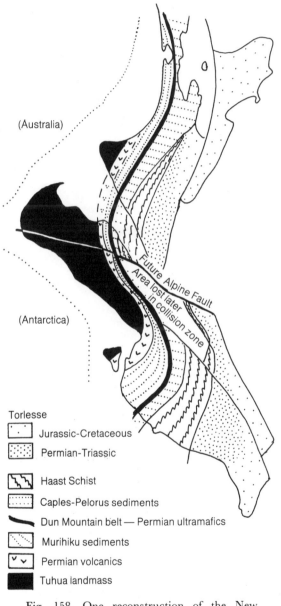

(Australia)

(Antarctica)

Future Alpine Fault
Area lost later in collision zone

Torlesse

- Jurassic-Cretaceous
- Permian-Triassic
- Haast Schist
- Caples-Pelorus sediments
- Dun Mountain belt — Permian ultramafics
- Murihiku sediments
- Permian volcanics
- Tuhua landmass

Fig. 158. One reconstruction of the New Zealand area about 130 million years ago (*after Korsch and Wellman*). Compare this with Fig. 83 on page 44, and see also Fig. 392 on page 197.

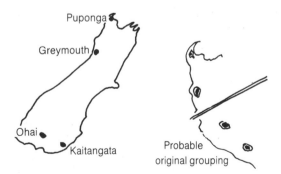

Puponga

Greymouth

Ohai

Kaitangata

Probable original grouping

Fig. 160. Areas of Cretaceous coalmines.

Coal is plant matter which has compacted and lost moisture and hydrocarbon gases. The nature of the plant material determines the kind of coal, since waxy spores or woody stems each make different types. The amount of alteration determines the rank of the coal (Fig. 161), based on the amount of moisture and gas driven off. It seems that the degree of alteration depends less on the age of the coal and more on the depth of burial and therefore the pressures and temperatures acting on the plant material. On the West Coast coal rank can vary along a single seam if it has been buried in a basin where the deepest burial occurred in the middle.

Generally, New Zealand coals are not of very high rank, being mostly bituminous or sub-bituminous. However, there was one small seam of high-grade anthracite in Canterbury, where volcanic rocks heated a coal seam for a short distance.

Between coal seams are layers of mud or quartz sands, left when rivers changed their courses or spilled over in floods. These layers of sediments are described as "coal measures" if coal occurs anywhere within them — even if the coal seams are missing in places, as they are in the coal measures by Thirteen Mile Stream, north of Greymouth.

Fig. 162. Cretaceous plant fossils from Berlins, Strongman and Rewanui.

The layers of sediment often preserve impressions of the plants that formed the coals (Fig. 162). At the beginning of the Cretaceous the forests included conifers, horsetails, cycads, seed-ferns and ferns, but during the Cretaceous the first flowering plants evolved. The Paparoa Coal Measures, at Greymouth, contain the oldest New Zealand coal seams, of middle Cretaceous age. On the dumps below the Strongman Mine (Fig. 163) mudstones can be found bearing fossil leaf impressions of ferns and early angiosperms, broad-leaved flowering plants.

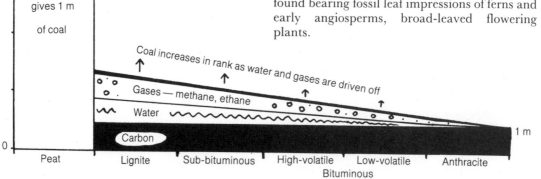

Fig. 161. The formation of coal.

Fig. 163. Coal measures in the valley below the Strongman Mine, with the black dumps barely visible in the foreground.

At Kaitangata, near the Clutha River outlet, and at Ohai in Southland the coal seams are late Cretaceous. There is only one set of coal measures at Kaitangata, but elsewhere all the first peat swamps seem to have been uplifted a little and eroded before a later series of swamp and flood plain deposits covered them and spread even further across the land in the Eocene Epoch, some 10 million years later.

In the opencast pits at Ohai, where the unwanted rock on top of the coal seams has been scraped off and stockpiled to one side, collectors can find plant remains in the red ironstone nodules. These are in the form of casts of wood and fossilised leaves of ferns and broad-leaved plants (Fig. 164). There are also sprays of the iron sulphide, pyrite, and shiny casts of the freshwater mussel, *Hyridella*.

So far, no remains of land animals with skeletons and no trace of reptiles has been found in these lowland sediments. They certainly existed, for our frogs, moa and tuatara must have had ancestors alive at that time, and one dinosaur vertebra has been found in marine sediments in Hawke's Bay.

In the millions of years during which the coal measures were forming the land behind them was being worn away. During the long weathering of the surface the minerals in the rocks broke down and the feldspars, especially, absorbed water and decomposed to form clays. Granites, with their high feldspar content, yield the best deposits of kaolin or china clay, but greywacke and schist also produce clay suitable for bricks and pipes; they are made red by the high iron content in these rocks.

The clays settled in low basins around the coal swamps, and many coalfields have brickworks associated with them. Near Kaitangata the red brick clays were used at Benhar (Fig. 165) from 1881, but today the clay pit to the west of the kilns is disused. For the present production of china at Benhar a whiter clay is

Fig. 164. Fossil freshwater mussels, *Hyridella* and an angiosperm leaf from the dumps at Ohai Opencast Mine.

Fig. 165. Clay pit at Benhar.

required; it is brought from coal measures near Charleston and mixed with imported clay.

While the clays and mica flakes were being washed out of the hills the remaining iron minerals decayed and only the quartz grains remained intact. (Quartz atoms are so tightly bonded that the mineral cannot decompose or be worn down easily.) Finally, the rolling landscape was mantled with sandy, leached soils and pebbles or grits of quartz. Today they survive as quartz sands in coal measures or cemented into massive quartzites and pebbly conglomerates.

The quartz conglomerates were important to gold prospectors. Since gold is an equally durable mineral and heavier than quartz, it remained trapped in river-laid (alluvial) deposits in areas with suitable source rocks. In Reefton the "cements" were mined at Crushington and up Lankey Creek where the remains of batteries can still be seen beside Devonian fossils. In Otago the first major goldfield was discovered when Gabriel Read panned gold in the gully eroding from the Blue Spur Conglomerate, which is coarse quartz and schist pebbles with a blue-grey cement (Plate 5G). Cliffs of the conglomerate surround the Goldfields Park picnic area up Gabriel's Gully, 3 kilometres from Lawrence.

Some of the Cretaceous land was eventually worn down to a nearly level surface or peneplain (from the Latin "paene", meaning almost). Later, much of this peneplain was covered with softer sediments, but in some areas these younger rocks have been eroded away and the older surface can be seen — on the broad uplands of Central Otago, on the slopes south-west of Pigroot Road inland from Palmerston, on the Gouland Downs on the Heaphy Track and on the plains above the Aorere and Takaka Rivers. The slope of the old peneplain can also be seen from the track to the Aorangi Mine, looking north past the coalfields of Mangarakau to Westhaven Inlet (Fig. 166). Visible in the bank on the right are thin coal seams, clays and sandstones of the Cretaceous-Eocene coal measures which were laid down on the peneplain surface.

Fig. 166. The sloping peneplain south of Mangarakau, on the west coast of North-west Nelson.

Fig. 167. New Zealand begins to split from the Gondwana landmass.

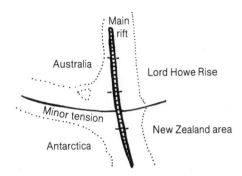

THE CRETACEOUS VOLCANICS

During the long weathering of the land subcrustal activity continued, but after the big collision of the Rangitata Orogeny the nature of the movement changed. A new upwelling and rift developed in the crust to the west of the New Zealand area, eventually breaking apart this edge of Gondwana and splitting us from the Australia-Antarctica block (Figs. 167 and 168.)

Although the main split was to the west, with fresh basaltic material rising to create the floor of the new Tasman Sea, there was considerable tension in other parts of New Zealand and Australia. The tearing apart of the crust created rift zones and fractures through which molten material was able to rise.

About 95 million years ago volcanoes began to erupt in Canterbury, from the Rangitata to

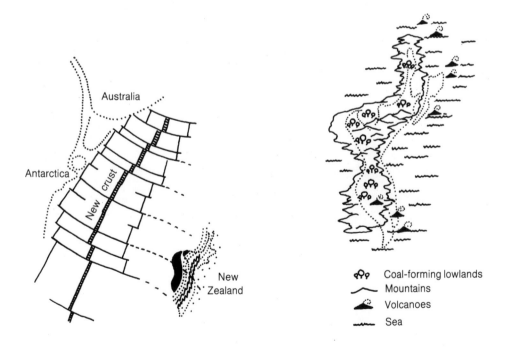

Fig. 168. Left, the Tasman Sea floor forming 85-55 million years ago. Right, a pattern of land and sea in Cretaceous New Zealand.

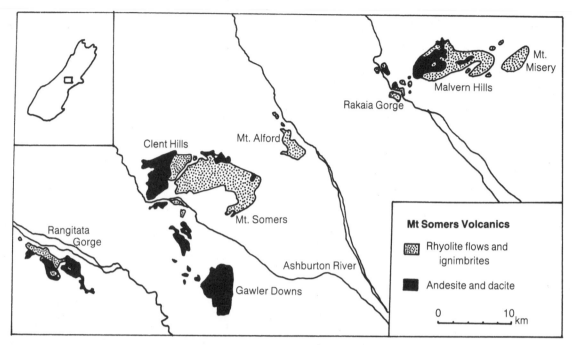

Fig. 169. Cretaceous volcanics in Canterbury (*after Adams and Oliver*).

the Hororata Rivers, producing the Mount Somers Volcanics (Fig. 169), which contain both rhyolites and andesites. Generally, the andesites erupted first, but not always. Here there are also dacite, a lava between rhyolite and andesite, and pitchstone, a kind of obsidian.

In the Rakaia Gorge (Fig. 170) the cliffs on the right facing upstream from the bridges are of andesite, and on the left they are mostly of rhyolite, containing tiny (3-millimetre) garnets and hexagonal quartz crystals.

The Mount Somers Volcanics have drawn rockhounds from all over New Zealand for the gemstones and crystals they contain. Gas bub-

Fig. 170. The Rakaia River gorge.

bles in the lavas left holes which later filled with crystals deposited by percolating groundwaters, heated perhaps by the remnants of volcanic activity. Some of the crystals are varieties of quartz (Fig. 171). Rarely, the crystals are large and purple — good amethyst (Plate 3I); more often they are smaller, white or water-clear, or even a pale green. Clear quartz crystals have been called Mount Somers Diamonds.

Some of the quartz here grew in layers of crystals so fine that they can hardly be seen, even with a microscope — up to 500 layers to the millimetre. This cryptocrystalline quartz (meaning hidden crystal) is called chalcedony when it is plain, agate when it is banded, and moss agate when it contains mineral growth in various colours (Plate 3F). Agates can still be found in cavities and seams in the andesite. (Some farmers in Canterbury allow collecting, while others charge a fee, and some find that intensive farming and rock hunters do not mix. Collectors must always ask permission to collect on farm land.)

During the millions of years these volcanic rocks have been exposed to the weather they have gradually broken down. Agates and other quartz materials resist weathering so they roll down the cliffs like so many potatoes, gradually making their way down to the rivers and eventually out to sea. There the waves tumble them along the shoreline and many find their way northwards along the coast until they are stopped by the barrier of Banks Peninsula.

Anywhere on this route agates may be found. On the smaller rivers like the Hinds, Hororata and Stour the proportion of agates may be higher than on the beds of rivers like the Rangitata and Rakaia, which carry large loads of greywacke from the alps; but luck is part of the search. Beaches in the area, especially Birdlings Flat, are usually rewarding (Fig. 172).

Quartz is not the only mineral to fill the cavities in these lavas. Calcite crystals grew in some, especially at Rakaia. Where this is the case agate sometimes fills the rest of the cavity, and on the outsides of the agate nodules are calcite crystals (Fig. 173) or merely the shapes of crystals that have since been dissolved by weathering.

Other common minerals to be found in cavities in volcanic rocks are the zeolites, so named because they contain water within their molecules which can be boiled off (from the

Fig. 171. Quartz crystals and banded agate, Mount Somers.

Fig. 172. Beach agates from Birdlings Flat.

Fig. 173. Agate and quartz grown inside a layer of pointed calcite crystals, Rakaia Gorge.

91

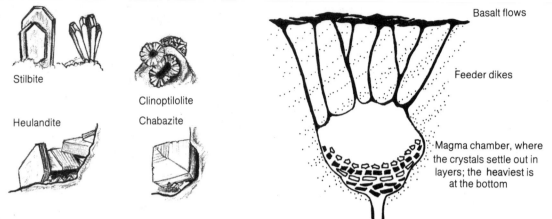

Fig. 174. Zeolites found in the Mount Somers Volcanics.

Fig. 175. Lava escaping from a magma chamber as it did in Marlborough.

Greek "zein" meaning to boil). Agates are sometimes found filling the cavities around zeolites, too. Zeolites form a large and confusing group, but some have distinctive forms (Fig. 174).

Stilbite is a pink or white zeolite, with pearly, kite-shaped crystals which sometimes completely fill the cavities. Clinoptilolite forms fan-shaped rosettes, also salmon-pink or white, at Rakaia Gorge. Heulandite is roughly coffin-shaped, and forms a white, spiky crust lining the cavities. Analcite forms garnet-like, white crystals with kite-shaped, flat faces on the rounded crystals. Chabazite is clear and almost cubic in shape.

Between Mount Somers and the Hororata River, trees were buried in volcanic ash and petrified or turned to stone (Plate 3A). This generally happens when wood is buried before it can rot and when the groundwaters seeping through the wood contain enough dissolved minerals to fill the wood cells. Sometimes the cell structure is preserved completely and at other times only a cast of the tree or limb is left, with agate or other minerals replacing the wood. Silica or quartz replacement can produce petrified wood for cutting and polishing; but the stones produced by the replacement of wood cells with calcite, pyrite and blue iron phosphate, vivianite, are not suitable for polishing.

At Petrifying Gully on Mount Somers opalised and agatised wood is found in a slip where the volcanic ash has turned to bentonite, a white clay which is gluey when wet and concrete hard when dry. The petrified wood here is often shattered, but at its best the wood grain is perfectly preserved in green, gold, orange, red and all shades of brown. It is very hard to find, and access very much depends upon farm activities at the time.

The volcanics further north are mostly basaltic in nature.

In Marlborough there were two major eruptions in the Awatere and Clarence Valleys. The molten lava rose up through feeder channels (dikes* that cut across the layers of older rocks like chainsaw blades) and erupted onto the land and nearby sea (Fig. 175), so the layers of ash and lava in the Awatere Valley contain plant and marine fossils that date them as Cretaceous (Fig. 176).

Deep underground in the lava reservoir the remaining magma began to crystallise olivines and pyroxenes, which separated into layers, making a large, layered intrusion of mafic and ultramafic rocks. In two localities these have been lifted up and exposed to view by later erosion. One intrusion forms Tapuaenuku, the highest mountain in the Inland Kaikouras, and the other Blue Mountain, further north. The mafic rocks from Tapuaenuku, especially syenite, may be collected in the Hodder River. At one time the syenite here contained large

*This spelling is preferred by the Geological Survey.

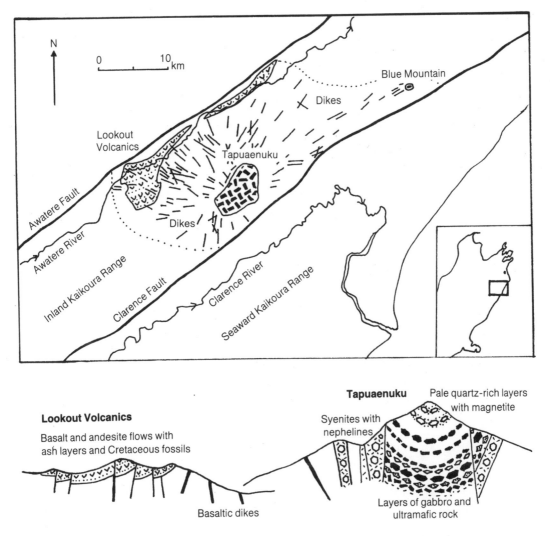

Lookout Volcanics

Basalt and andesite flows with
ash layers and Cretaceous fossils

Basaltic dikes

Tapuaenuku

Pale quartz-rich layers
with magnetite

Syenites with
nephelines

Layers of gabbro and
ultramafic rock

Fig. 176. Geological map and cross-sections of the Cretaceous volcanics in Marlborough (*after Nicol*).

crystals of nepheline, a silicate related to the feldspars but with a lower proportion of silica; but the nepheline has altered to a fine-grained intergrowth of the zeolite, analcite, and white mica. Only the nepheline shape remains. (Such replacement of one mineral by another, keeping the shape of the first mineral, is called a pseudomorph, meaning false shape.) See Fig. 177.

In North Canterbury syenite without nepheline makes up the bulk of Hurunui Peak, to the north of Balmoral Forest. Its pale feldspars spotted with dark-green hornblende and black biotite mica can be seen in the Mandamus River

Fig. 177. Nepheline pseudomorphs in syenite, Hodder River.

Fig. 178. Above, syenite in the Mandamus Gorge. Below, syenite dike with dark chilled margin.

gorge by Tekoa (Fig. 178), where the peak is cut by dikes of fine-grained, black lava showing a chilled margin on each side where the lava cooled quickly against the surrounding rock (Fig. 179). Where the syenite butts against the surrounding Torlesse greywackes it is so fine-grained that at first glance it looks just like sandstone.

There is a mass of gabbro exposed in a small stream to the north-east between Hurunui Peak and the Pahau River. Dikes radiate from both plutonic bodies, the syenite of the Mandamus and the gabbro of the Hurunui-Pahau area, for some 5 kilometres in exactly the same way as the dikes swarm around the Tapuaenuku centre.

In the North Island basaltic eruptions made the Tangihuas in Northland (page 144) and the very similar Matakaoa Volcanics at the top of the East Coast from Cape Runaway to Matakaoa Point. The rocks are basalt, gabbro and volcanic debris, which on the beaches by Lottin Point contain zeolites and small agates. The volcanics also form the uplands south-west of Hicks Bay, and grey agates from them can be found on the beaches behind the Te Araroa Motor Camp and among the rocks before Onepoto Bay. Here pillow lavas have fossils of the Oligocene Epoch among them, especially large oysters, which led earlier geologists to believe that the lavas were Oligocene, like the pillow lavas at Oamaru (page 121). However, there are Cretaceous forams in the sediments with other Matakaoa Volcanics and we now believe the Onepoto shells were later deposited on the lavas much as the shells of today rest among these rocks.

Fig. 179 Matakaoa Volcanics. Above, Onepoto Bay in the foreground and Matakaoa Point beyond Hicks Bay. Left, Oligocene conglomerate on top of pillow lavas. Below, fossil oyster in the conglomerate.

SEDIMENTS OF THE CRETACEOUS SEAS

The presence of pillow lavas shows that the volcanoes were undersea. All the North Island Cretaceous volcanics seem to have been erupted on to the sea floor on the north-east side of the Cretaceous land mass, where another offshore slope was being covered with the same kind of sediments as those that produced the earlier Torlesse rocks.

In fact, the oldest marine Cretaceous rocks are so similar to those of the Torlesse that some of the rocks shown as Cretaceous on older maps are now proving to be late Jurassic Torlesse, and the relationships between the two all along the east coast, in North Canterbury and Marlborough and in the ranges north-west of Gisborne where most of the Cretaceous type sections are found are still being studied.

The fossils used to subdivide the Cretaceous rocks into series and stages are mainly species of the bivalve *Inoceramus* (pronounced "I know Sarah Muss"), or *Ino* for short (Fig. 180). From the middle Jurassic these shellfish evolved a series of species which varied in size and shape and never repeated themselves. Like the earlier Permian *Atomodesma*, the *Ino's* shells were made up of cross-grained fibres which sometimes fill the rocks; some of the shells were up to 600 millimetres across, and the prismatic calcite fibres of *I. rangatira* are up to a centimetre long. *Aucellina*, a rounded bivalve shellfish the size of a walnut, is another important early Cretaceous fossil in New Zealand. The cephalopods (Fig. 181) are represented by belemnites, baculites (a straight ammonite) and a limited variety of

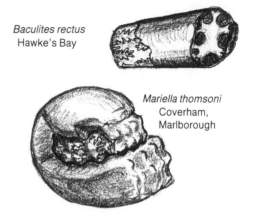

Baculites rectus
Hawke's Bay

Mariella thomsoni
Coverham,
Marlborough

Fig. 181. Mata Series ammonites.

round ammonites which developed some unusual coiling patterns before becoming extinct with all the other ammonites and belemnites at the end of the Cretaceous Period.

The oldest Cretaceous series is named after Mount Taitai in the Raukumara Ranges (Fig. 182), which, like the other mountains close by, stands up from the softer mudstones around it as an isolated mass of hard sandstone containing no fossils to give it a date. See Fig. 183.

The oldest Cretaceous fossils are found to the south-west in the Koranga Valley off the Upper Waioeka River, where layers of sandstone and conglomerate contain fossils which define the Korangan Stage and show the stages of the Clarence Series as well. The Clarence Series is named after the river in Marlborough where there are fine examples of Cretaceous rocks in

Fig. 182. Lower Cretaceous sandstone mountains. From left, Aorangi, Taitai and Wharekia.

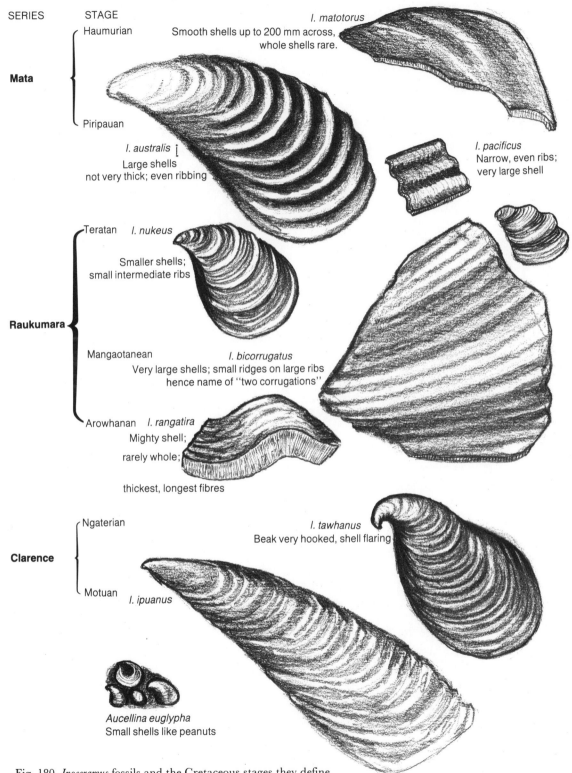

SERIES

STAGE

Mata

Haumurian

Piripauan

I. matotorus
Smooth shells up to 200 mm across,
whole shells rare.

I. australis
Large shells
not very thick; even ribbing

I. pacificus
Narrow, even ribs;
very large shell

Raukumara

Teratan   *I. nukeus*

Smaller shells;
small intermediate ribs

Mangaotanean   *I. bicorrugatus*
Very large shells; small ridges on large ribs
hence name of ''two corrugations''

Arowhanan   *I. rangatira*
Mighty shell;

rarely whole;

thickest, longest fibres

Clarence

Ngaterian

I. tawhanus
Beak very hooked, shell flaring

Motuan   *I. ipuanus*

Aucellina euglypha
Small shells like peanuts

Fig. 180. *Inoceramus* fossils and the Cretaceous stages they define.

97

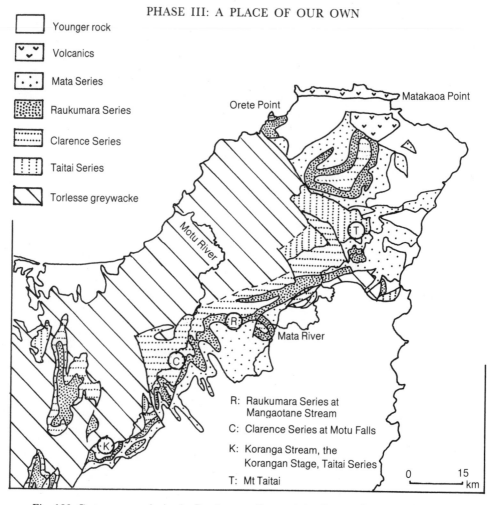

Younger rock

Volcanics

Mata Series

Raukumara Series

Clarence Series

Taitai Series

Torlesse greywacke

R: Raukumara Series at
   Mangaotane Stream

C: Clarence Series at Motu Falls

K: Koranga Stream, the
   Korangan Stage, Taitai Series

T: Mt Taitai

Fig. 183. Cretaceous rocks in the Raukumara Range (*after Suggate, Stevens and Te Punga*).

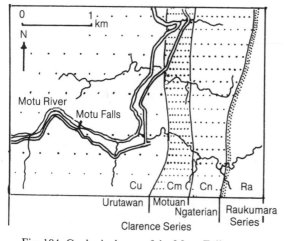

Fig. 184. Geological map of the Motu Falls area
(*after Suggate, Stevens and Te Punga*).

the Coverham area at the end of a dusty and winding private farm road.

The Clarence stages were described and named in the Motu Falls area, some 3 kilometres east of Motu village, with names from the Urutawa and Ngateretere survey districts (Fig. 184). There is a viewing place at the falls, and in the next kilometre downstream the river crosses layers of sandstone and mudstone. Access is difficult and the casual visitor would see few fossils. There is a layer of concretions with some fossils in them at the beginning of a big roadcut about a kilometre east of the falls; this roadcut is a good example of the early Cretaceous rocks with their rusty weathering (Fig. 185).

Just as remote and on a Forestry track are the type sections for the Raukumara Series, under

Fig. 185. Roadcut east of Motu Falls, with concretions in the layers on the right.

Arowhana Mountain in the Mangaotane and Te Rata Streams.

The Mata Series is named after the river which flows for most of its length through rocks of this age, but the type sections and stage names come from the South Island at Amuri Bluff just south of Kaikoura (page 104).

Travellers in the East Cape area who want to see a good Cretaceous sequence could walk the shore platform from Waihau Bay to Orete Point (Fig. 186 and 187); the land is private property, so care must be taken with the tides.

A fault runs through Waihau Bay, and the rocks to the east are older, going back to the Clarence Series. The rocks on the shore to Orete Point have all been overturned, so although you seem to be walking up the layers you are actually going through them from youngest to

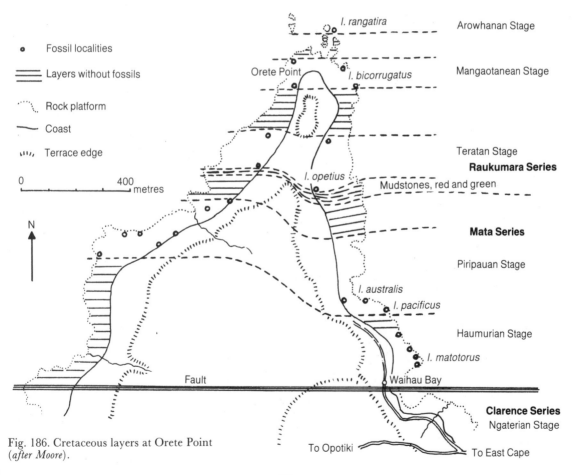

Fig. 186. Cretaceous layers at Orete Point (*after Moore*).

99

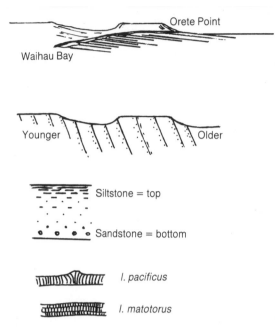

Fig. 187. Geology at Orete Point. Top, general view. Centre, top, the rock layers are overturned. Centre, bottom, graded bed formed from one flow of material with the coarsest settling out first. Bottom, cross-sections of *Inoceramus* shells.

Fig. 188. The mudstones just west of the red and green mudstones contain fragments of *Inoceramus opetius* shells on edge, which look like hard earthworms sticking on the rock. The photograph shows their location.

oldest (Fig. 187). Graded bedding is visible in some of the layers of sandstone and siltstone — the layers grade up from coarse sandstone at the base to muddy siltstone at the top, like the layers in the Torlesse greywackes (page 66); this shows the original placement of the layers.

Most of the fossils present here are just pieces, but they occur in bands which can be traced the whole width of the beach.

The youngest Cretaceous beds, visible by Waihau Bay, are of sandstones with graded beds and plant fragments, some lying in ripples left by currents on the Cretaceous sea floor. The shells of *Inoceramus matotorus* and *I. pacificus* (Fig. 187) are present in scattered fragments; they can be distinguished by whether or not there are layers in the fibres.

The conspicuous red and green mudstone layers contain no fossils, but just beyond them pieces of *I. opetius* are scattered through the silty mudstone beds (Fig. 188). Beyond the barren zone by Whaka Island pieces of *I. bicorrugatus* are also widely dispersed. Pieces of *I. rangatira* from the oldest rocks in the area were found in one concretion at the end of the rocks at low tide. They are also found on the beach by Raukokore church.

Although the earliest Cretaceous rocks are as hard as the Torlesse greywackes, the later Cretaceous rocks are much softer and can easily be broken by a hammer blow in the same way as the Cenozoic or Tertiary rocks which lie on top of them.

Cretaceous sediments further south

The softer rocks of the Raukumara and Mata Series can be seen in many places on the eastern side of New Zealand from Hawke's Bay to Canterbury.

The Cretaceous rocks in the hills along the coast from Hastings south are mostly blue-grey siltstones and a softer, blue-grey mudstone with bands of limestone containing *Inoceramus* prisms and sometimes complete shells, together with very rare ammonites and belemnites.

On the coast at Red Island (Fig. 189), south of Waimarama, pink limestone contains some *Aucellina* shells in layers beside the pillow lavas, another of those Cretaceous volcanics. Among the volcanic rocks are some jaspers and chert, colourful forms of silica which are opaque and here often cracked. More noticeable are the zeolites thomsonite and natrolite (Fig. 190),

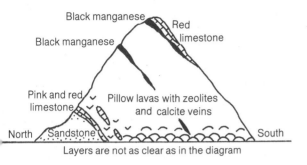

Fig. 189. Geology of Red Island.

Fig. 191. Photograph of concretions on the beach south of Red Island and drawings of their structure.

found as white, radiating masses between chunks of lava among the rocks on the south side. Red Island is a tiring walk away down the beach at low tide, or a steep climb over farmland, and at high tide waves cover the spit of sand that joins it to the mainland.

The Red Island rocks wash up on the beach at Kairakau, which can be reached by road. At Mangakuri Beach, next south down the coast, a walk of some distance south down the beach brings collectors to boulders of somewhat porous red and yellow jasper and even pale amethyst with inclusions of copper and iron pyrites. Just at the end of the sandy beach the Cretaceous rocks are present in the form of layers of sandstones and siltstones with some

conglomerates, originally deposited in shallow water. They contain cross-beds outlined in fine grey and black bands (Plate 4A), the black being plant fragments which washed into ripples on the sandy sea floor just as ironsands lie in ripples on beaches today. In the beds are nodules of iron pyrite, which can also be found at Waimarama. Some of the siltstones contain *Inoceramus* fragments and a few whole shells, and there are concretions by the dozen here and in many of the younger mudstones.

Concretions (Fig. 191) are rounded masses of hardened rock which formed soon after sediments were deposited on the sea floor. Small patches of organic matter or other mineral material attracted cementing minerals and gradually spheres or cylinders of harder rock built out into the surrounding sediments, cementing the grains together without pushing them aside. Some may have had a core of soft material which left no trace, some contain worm tubes, casts or dung in a shapeless mass and others contain fossils such as crabs, ammonites and other shells.

Fig. 190. Natrolite from Red Island.

101

Sometimes, after forming into hard spheres, the concretions developed cracks in the core which were then filled with crystals deposited by percolating groundwaters. These are known as septarian concretions. The deposited crystals are usually of calcite, but at Mangakuri they are also of barite (barium sulphate), sometimes massive and sometimes forming clear, golden crystals (Plate 4D). The barite crystals are often enclosed in white calcite, which can be etched away with dilute acid.

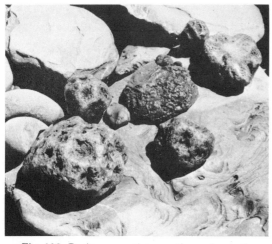

Fig. 192. Barite concretions on the rocks south of Porangahau Beach.

Fig. 193. Animal trails on sandy Whangai Shale at Kate's Quarry, Porangahau.

Barite is also found on the beach south of Porangahau (Fig. 192), where the rocks are younger and the barite forms the whole of a grey, radiating nodule, distinguished by its extra weight because barite is a heavy mineral.

On the Whangai Range, east of Dannevirke, the blue-grey mudstones are covered by a pale, chocolate-coloured, very fine mudstone — the Whangai Shale, Haumurian in age. This weathers to a distinctive pinkish-white colour and sometimes has a yellow bloom of jarosite (a sulphur mineral) on the weathered rock. These shales are common along the southern east coast and in Marlborough and can be seen at Kate's Quarry near Porangahau, on the highway west of the Wimbledon turnoff. Some of the layers are sandy, and animal trails are visible on the surface of the beds (Fig. 193).

In the upper Cretaceous rocks of Hawke's Bay/Marlborough there are also beds of greensand, coloured by grains of dark-green glauconite, an iron mineral which forms on a sea floor where very little sediment is accumulating from the land (Plate 5A). Glauconite sands show that the nearby land was low and slowly eroding, but they do not show the depth of the sea in that area as they form today at depths varying from a few metres down to several hundred metres.

Cretaceous sediments continue all down the southern North Island east coast. East of Masterton, at Ngahape in the Ngaumu State Forest, there are upper Cretaceous sandstones, mudstones and shales with the occasional *Inoceramus* (these are more plentiful in the Little Kaiwhata River) and rare belemnites and ammonites. On the hills above the Totara Stream (Fig. 194) are pillow lavas which contain agates, some quartz (occasionally pale amethyst in colour) and some zeolites, especially analcite (Fig. 195 and 196). Poor-quality petrified wood can also be found in the Kaiwhata River,

Fig. 195. Blocky analcite and radiating datolite on banded calcite, from Totara Stream.

Fig. 196. Pale amethyst quartz in agate, Totara Stream.

Fig. 197. *Inoceramus* plates and belemnite, Tora Beach, Wairarapa.

together with occasional quartz and agate pebbles, but there has been so much wrenching of the Wairarapa area that most of the agates here are badly cracked.

Further south, the Aorangi Mountains are mostly comprised of harder Jurassic and Cretaceous greywackes, but on their eastern side the younger Cretaceous sediments contain *Inoceramus* fragments and shells, belemnites and tiny, black scraps of fossilised plant material (Fig. 197).

Some of the Cretaceous concretions contain

bone — the remains of marine reptiles related to dinosaurs. In one remote area of inland Hawke's Bay enough bones have been collected to identify whole animals — a new species of mosasaur (Fig 198), a bony fish and a turtle. All these bones have been carried out in backpacks, laboriously dissolved in acid and then identified by one amateur worker, who is now becoming

Fig. 198. Mosasaur jawbone and sketch of this lizard-like marine reptile.

Fig. 194. Seaview Road in Ngaumu State Forest crosses Totara Stream before climbing up to outcrops of pillow lavas and zeolites on the summit of Centre Trig.

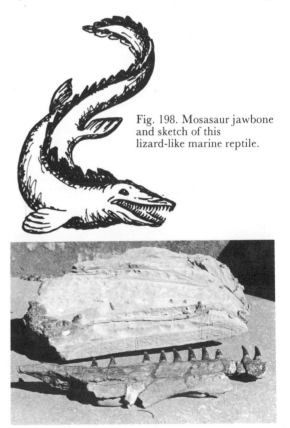

Fig. 202. Sketch of plesiosaur.

Fig. 199. Vertebra of a therapod dinosaur about 4 metres high.

an expert in the field. One single vertebra (Fig. 199), from a concretion which formed on the sea floor, was found to be that of an upright carnivorous land dinosaur which was perhaps washed out to sea during a flood. So far it is our only dinosaur fossil.

Concretions containing bone also occur in the South Island, further down the Cretaceous seaway. At Oaro, just south of Kaikoura, the sea is eroding a bank of late Cretaceous rock just north of Amuri (or Haumuri) Bluff (Fig. 200), and these rocks have become the type section for the rocks of the Mata Series, the Piripauan and Haumurian.

As you walk south on the track beside the railway line from Oaro, there is a long, sandy beach which ends in a boulder-strewn area. Among these boulders are some containing bone and others with masses of belemnites and some sharks' teeth (Fig. 201). The layers of belemnites are reminiscent of Jacques Cousteau's photographs of the sea floor after the mating and death of squids; this could be an explanation for the mass destruction represented by these rocks.

The bone found here generally belonged to marine reptiles such as mosasaurs and plesiosaurs (Fig. 202). Water-worn vertebrae are relatively common, and south of the tunnel there are very large concretions containing groups of bones (Fig. 203 and Plate 4F).

A magnificent specimen recently collected contained a head with rows of 50-millimetre

Fig. 201. Layers of belemnites in rocks at Amuri Bluff.

Fig. 203. Concretions showing reptile bones, Amuri Bluff.

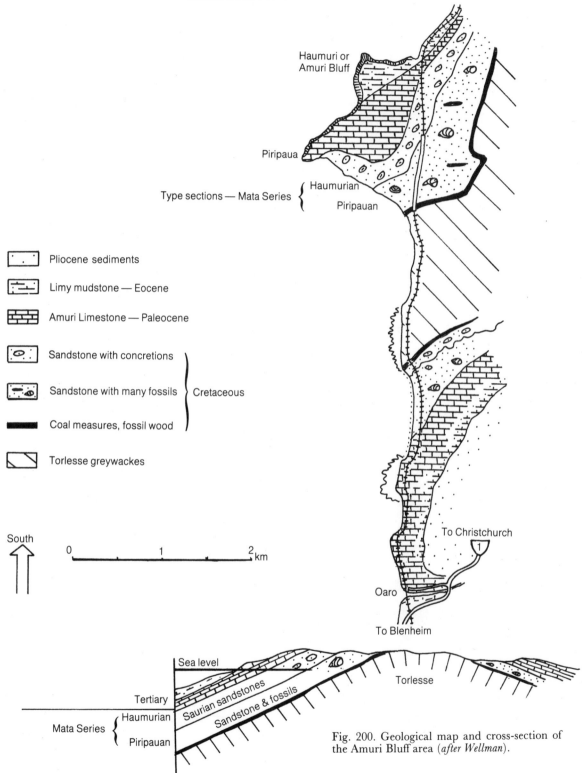

Haumuri or
Amuri Bluff

Piripaua

Type sections — Mata Series { Haumurian

Piripauan

Pliocene sediments

Limy mudstone — Eocene

Amuri Limestone — Paleocene

Sandstone with concretions

Sandstone with many fossils } Cretaceous

Coal measures, fossil wood

Torlesse greywackes

South

0   1   2 km

To Christchurch

1

Oaro

To Blenheim

Sea level

Tertiary

Mata Series { Haumurian

Piripauan

Saurian sandstones

Sandstone & fossils

Torlesse

Fig. 200. Geological map and cross-section of
the Amuri Bluff area (*after Wellman*).

105

teeth; it is now in the Canterbury Museum. Fossils such as these are rare and must not be damaged. If they can be taken from the beach without breakage they may be removed, but they must be shown to people at a university or the Geological Survey for listing in the Fossil Record File. Private collectors holding such material must take good care of it, and if at any stage the owner loses interest the fossil should go to the Geological Survey. If the fossil is firmly embedded in a cliff or large concretion it must be left for specialists to remove. Report it to the nearest university geology department or Geological Survey branch.

At Oaro beach there is also petrified wood (Fig. 204) — some pieces containing iron pyrites; others with unusual calcite growths and a few pieces containing enough agate that they polish well. These are eroding from the coal measures at the south end of the first beach and up the stream.

On the sea cliffs just past the point where the railway line curves into a tunnel and a steep track leads down to the beach, sulphur forms yellow crystals on the bank. Near here the sandstones contain crystals of gypsum, which is sometimes called selenite. This mineral, calcium sulphate, is visible as shining rosettes or long crystals, or even as sand-filled, flower-like structures. It formed as the gypsum crystals grew in the sandy sediments and cemented the sand together to make a sand selenite. The crystals have good cleavage and can be scratched with a fingernail.

The white limetone of the Amuri Bluff itself is a unique deposit which occurs from North

Fig. 205. Forams in the Amuri Limestone as seen under a microscope.

Canterbury to the Wairarapa. It was formed from the shells of a multitude of tiny, one-celled creatures, whose calcite shells (Fig. 205) accumulated to form a chalk ooze on the bottom of a probably warm and quiet sea, the depth of which is uncertain. The conditions that produced this hard, white or creamy limestone persisted for a long time, which means the material may be of different ages in different places. Its time of origin may vary from the end of the Cretaceous up to the Oliogocene; at the Amuri Bluff the limestone is Paleocene.

Travellers notice this limestone especially around Kaikoura and the hills to the north and south. At Kaikoura Peninsula and many other places it contains nodules or lumps of flint, a massive form of silica. Among the tiny creatures whose remains formed the chalk ooze were diatoms (Fig. 206) and radiolarians, with shells of silica constructed from the amorphous (non-crystalline) common opal. The silica from these shells dissolved in the sediments, then reformed as flint nodules.

The flints vary in colour from pale to dark grey, brownish and even red or pink. In some places the silica was confined to spaces between the layers of sediment to make a rock called zebrastone (Fig. 207), with layers of opaque, white limestone and darker, translucent flint.

Fig. 204. Petrified wood with calcite rind, Oaro.

Fig. 206. Diatoms.

Fig. 207. Layers of silica and limestone known as zebrastone.

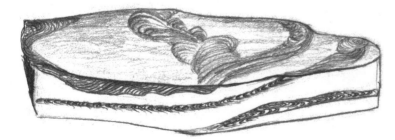

Fig. 209. Sponge fossils at Chancet Rocks.

Zebrastone may be found from Hawke's Bay to Kaikoura, although it is rarer in the north.

Flint nodules are fairly common in chalky limestone overseas. The chalk cliffs of Dover are also Cretaceous in age (remember, Cretaceous comes from the word "creta" meaning chalk) and are studded with flints which weather out to form the beaches below — and very hard, uncomfortable beaches they are. Flints were often used as ballast in sailing ships and can be still found dumped on beaches in our main harbours; for example, in Wellington at Balaena Bay.

Large fossils are rare in these limestones, although a few sharks' teeth have been found. Most common are trace fossils, visible in the form of rows of grey arrowheads on the broken rock or swirling patterns on the surface (Fig. 208). These are the feeding traces of a wormlike

creature, possibly something like a sea-pen, which extended from a central burrow to sweep across the sea floor for minute scraps of food.

However, some spectacular fossils can be seen in the white limestone at Ward Beach, 3 kilometres east of Ward in Marlborough. At the Chancet Rocks, about 1 kilometre north along the beach and accessible only by walking as the road is blocked off, there are beds of fossilised sponges (Fig. 209). These sponges probably flourished on the quiet sea floor until an influx of sediment buried them; sometimes they were able to grow again, the new head rising from the core of the old one, but more often the layers of sponges are separate. In the millennia since the sponges grew on the sea floor in layers the rocks have been tilted and now the sponges hang on the surfaces of the limestone like clocks on a wall. They show none of the original texture, just a cast of general sponge shape. At Chancet Rocks they are protected, but at the quarry on the way to the beach collectors may be allowed to search for specimens.

Fig. 208. Feeding traces in Amuri Limestone, and an animal that may have formed these traces, which are known as fucoid markings.

THE END OF AN ERA

South of the Chancet Rocks, Woodside Creek (Plate 4E) is one of the few places in New Zealand where geologists have found layers of rock that exactly mark the transition from the rocks of the Mesozoic Era to the rocks of the Cenozoic Era, the end of the Cretaceous Period and the beginning of the Tertiary. (Tertiary is an old term generally used by geologists to cover the period after the Cretaceous and before the Ice Age; the Quaternary period is the last 2 million years.)

The boundary is hardly marked by the sediments themselves. To a casual observer they all appear to be fine, creamy limestones, although a specialist in limestones may say they vary from a muddy limestone to a limy mudstone. The time change is instead marked by the evidence of fossils.

The Cretaceous Period ended when vast numbers of Cretaceous animals became extinct in one of the world's great mysteries. Dinosaurs, ammonites, belemnites, *Inoceramus* and whole families of other bivalves disappeared. Everywhere, one-celled plankton were suddenly reduced to a few surviving species. These, together with a thin band of clay a few millimetres thick and containing a slightly higher-than-usual amount of rare earth elements, mark the boundary in Woodside Creek (Fig. 210). (The unusual concentration here of elements such as iridium has led some scientists to consider cosmic fallout or the impact of a giant meteorite or comet as a cause for the extinctions.) Whatever the reason, in one layer of rock the tiny animals are packed in and typical of the rich variety of Cretaceous forms, and in the rock some 12 centimetres higher only a few creatures are left, including some of the new species that evolved from the survivors.

The transition to the Cenozoic Era can also be seen in the rocks of the Te Uri Stream in Hawke's Bay and in Waipara, North Canterbury (see page 112). In all three locations — Woodside Creek, Te Uri Stream and Waipara — the passage of time is shown by microscopic fossils, mostly Foraminifera.

Foraminifera are commonly called forams and have been known for a long time, but they were not the subject of detailed study until oil exploration became big business. Most oil drilling is done with a drill bit with three rotating

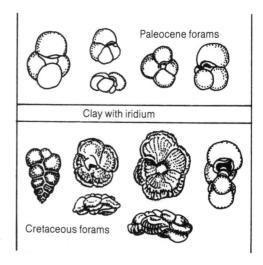

Fig. 210. The Cretaceous-Tertiary boundary at Woodside Creek lies in the thin layer of clay (arrowed). (The holes were drilled by geologists taking samples for study.) The diagram shows the change in the forams in the rock. (*From drawings supplied by P. Strong, N.Z.G.S.*)

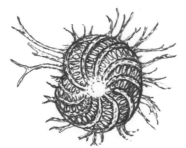

Fig. 211.
Foraminifer.

THE NEW ERA
The Dannevirke Series — from the Paleocene to the Eocene Epochs

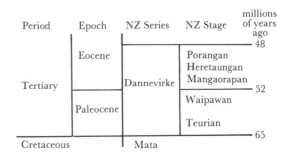

Period	Epoch	NZ Series	NZ Stage	millions of years ago
				48
Tertiary	Eocene	Dannevirke	Porangan	
			Heretaungan	
			Mangaorapan	
				52
	Paleocene		Waipawan	
			Teurian	
				65
Cretaceous		Mata		

heads. These chew the rock into tiny fragments which come up the well-hole mixed with a greasy mud which is used as a lubricant. The washed chips of rock are all we have to date the rocks down the hole, and for this purpose we need fossils small enough to survive intact in a 2-millimetre chip. Such fossils are mostly the remains of single-celled plants and animals.

Forams (Fig. 211) are animals that generally secrete a calcite shell with a number of openings (foramina) through which the animal can poke out streamers (pseudopodia) to catch even smaller creatures for food.

Forams are widespread in the oceans, and a few are as large as sand grains. Since they exist today they can be sorted from modern sands with a strong magnifying glass or microscope. They have evolved steadily throughout their history and have been so well studied that a sample of rock with forams can be as easily dated as a rock containing a fossilised dinosaur. In New Zealand forams have been very useful in dating the thick layers of siltstones and mudstones that contain no larger fossils.

After the Cretaceous the sediments coming onto the sea floor along the east coast became increasingly muddy and less sandy as the uplands wore down; the Torlesse sedimentary rocks were eroding to make even finer sediments. All these silts and muds made rather monotonous rocks, with beds of different ages looking very much the same. The only fossils present in any numbers in them are forams, which are not of much interest to the beginner or anyone else without fine-mesh sieves, a microscope and lots of patience.

However, it was only in this eastern seaway that marine beds were laid down in continuous sequence from the Cretaceous Period to the Paleocene Epoch, so the search for type sections led to the Dannevirke Subdivision near Porangahau and to the Te Uri (Fig. 212) and

Fig. 212. Te Uri Stream.

Mangaorapa Streams. Here, in the Whangai Range, the Whangai Shales which contain Cretaceous fossils also contain Teurian Stage or lowest Paleocene forams in their upper layers, indicating a boundary change just like that at Woodside Creek. On top of the Whangai Shales are very thick deposits of white or greenish-grey muds, which fill the valley to the east towards Porangahau. They contain the entire series of forams evolving right through the Paleocene and all the way into the Miocene.

Unfortunately, the beds in this area are slumped and very hard to see. The sequence is less complicated near Waipawa, so a standard reference section has been described from this area, covering 1.3 kilometres along Tikokino Road from the town. Even here contact with the Cretaceous rocks is obscure; geologists also had to use the hills for several kilometres to the north when sorting out the positioning of the layers (Fig. 213).

The Cretaceous rocks in this area are creamy-grey argillite with flinty beds near the top. The beds of the Teurian Stage are more colourful

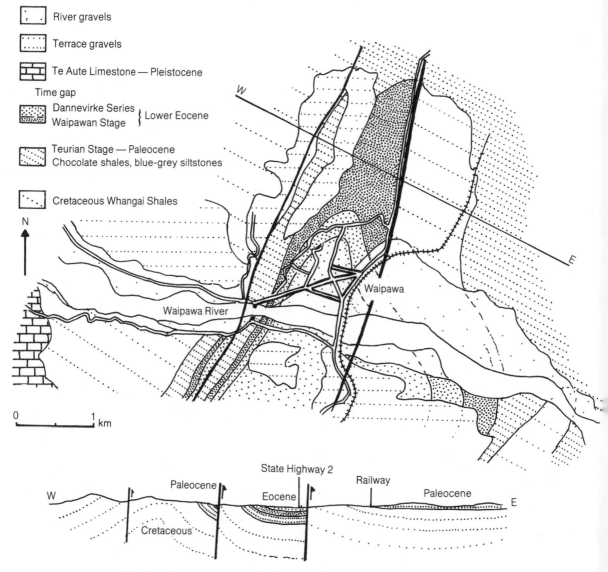

Fig. 213. Geological map of Waipawa and cross-section (*after Kingma*).

110

blue-grey siltstones changing to chocolate-brown with a yellow jarosite bloom. The chocolate-brown mudstones (Plate 4B) are known as the Waipawa Black Shales and were probably deposited in basins where little oxygen was available. The Waipawan Stage itself is east of the old Waipawa bridgehead. From here to the township all the stages of the Dannevirke Series are visible and were described from roadcut and river bank in massive mudstones which are paler as they decrease in age.

GOING UNDER — THE EARLY TERTIARY PERIOD: 65 TO 25 MILLION YEARS AGO

During the Early Tertiary the sea gradually moved inland across the South Island. At Oaro we have already seen how the Cretaceous land sediments containing petrified wood were covered by marine sands, then limestones. It has been mentioned that at Kaitangata in the south the coal seams were covered with marine sands, which at Wangaloa contain the best preserved earliest Tertiary (Teurian) fossil shells in New Zealand. All around New Zealand the same pattern can be seen. The land uplifted by the Rangitata Orogeny was being worn away and gradually covered by the sea. Something was going on.

It seems possible that our independence sank us. As long as New Zealand was part of the Gondwana landmass it was held up on the edge of the thick, continental mass of light material. When we split apart and moved away, the crust under this area was not as thick as continental crust usually is — it measured only some 26 kilometres compared with 37 kilometres. Having a thinner layer of buoyant crustal material, we sank lower in the mantle; and as we settled down the seas slowly rose over the land (Fig. 214). By the late Oligocene Epoch, about 25 million years ago, most of the old land area was

under the sea. Then in the Miocene renewed pressure and uplift caused the seas to retreat.

Some areas stayed out of the water longer than others, and these were sometimes the first to rise again, but in general the pattern was the same.

To help you see the rocks of each district more clearly we will go around the country looking at type sections and other places where the layers of rock show past events particularly well. In doing so we must focus not only on these rocks but also on the older and younger rocks that frame them.

North Canterbury

At Waipara the landscape records the passing of some 50 million years, during which time the sea covered the area to an increasing depth before gradually becoming shallow again. Up Georges Road, north from Amberley, to Ram Paddock Road and Laidmore Road the whole sequence of rocks is well exposed and has been visited by generations of students.

Even to the casual observer there is a difference between the jagged peaks from the Torlesse to the Tekoa Ranges and the more gently rolling landscape towards the sea, where the layers of younger rocks are draped over folded anticlines and synclines.

Looking at the layers of rocks near Laidmore (Plate 5H) we can see that the lowest western rocks are those of the basement Torlesse, crumpled, tilted and with an eroded and weathered upper surface, showing that the Torlesse had become a land that wore away to a lowland area, for on top of it lie layers of quartz sands and thin coal seams.

After these rocks were formed the sea moved in. Up the Waipara Gorge is one of the old collecting areas for Cretaceous marine reptiles, mosasaurs and plesiosaurs, which are now in the Canterbury Museum or lost overseas. (The first collection left in a sailing ship which dis-

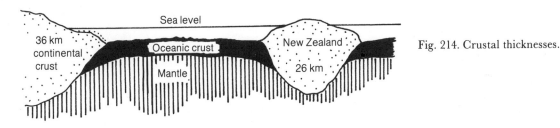

Fig. 214. Crustal thicknesses.

Fig. 215. Cross-section at Waipara.

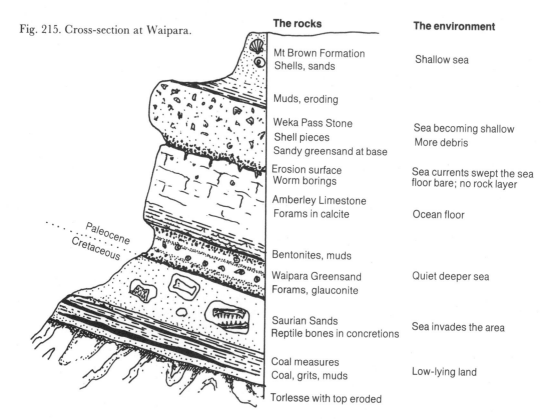

The rocks	The environment
Mt Brown Formation Shells, sands	Shallow sea
Muds, eroding	
Weka Pass Stone Shell pieces Sandy greensand at base	Sea becoming shallow More debris
Erosion surface Worm borings	Sea currents swept the sea floor bare; no rock layer
Amberley Limestone Forams in calcite	Ocean floor
Bentonites, muds	
Waipara Greensand Forams, glauconite	Quiet deeper sea
Saurian Sands Reptile bones in concretions	Sea invades the area
Coal measures Coal, grits, muds	Low-lying land
Torlesse with top eroded	

Paleocene
Cretaceous

appeared without trace; the second arrived in America, but has since been mislaid. The fossils were sent overseas for study and comparison with other saurians.) Two good collections have been made recently.

The cross-section (Fig. 215) shows how the layers have been interpreted, giving the history of the area from the time of the Cretaceous coal swamps some 70 million years ago to the last time a shallow sea covered this area, about 15 million years ago in the middle Miocene.

Equal thicknesses of rock do not represent equal lengths of time. Shallow-water shells and sands may build up quickly and represent a fairly short period of time; but greensands mark a time when little sediment was arriving from the land and build-up was slow, so the same thickness here represents a much longer period of time. An erosion surface shows the loss of a layer of unknown thickness. The shells of the forams which form the white, chalky limestones are so tiny that it takes a very long time for even a thin layer to form.

Fig. 216. At Weka Pass worm borings on the top of white Amberley Limestone are filled with greensand from the base of the overlying Weka Pass Stone.

112

These white, Oligocene limestones are found all over North Canterbury. At first sight they look very much like the older Amuri Limestone (page 106), even to their trace fossils, but the fossil forams in them show their difference in age. Both formations are important sources of lime for agriculture. The Amberley Limestone shown in the cross-section tends to be blocky, with many joints. To the west it becomes more chalky until near Oxford it is described as chalk. At Weka Pass, borings can be seen on the upper surface of the limestone (Fig. 216).

In most places the Weka Pass Stone in Fig. 215 is a massive, more resistant rock which overhangs the Amberley Limestone and contains the greensand glauconite.

Towards the sea all the Miocene layers are blue siltstones; only inland do they contain the Mount Brown Limestone beds shown at the top of the illustration.

The Arnold Series — late Eocene

West Coast

In the area just inland of Greymouth the layers of rock from the basement to those of the Miocene Epoch are draped over a vast, uplifted block, around which some of the younger material has been worn away to show the structure of the land. Of course, it has been sculptured by erosion, broken by small faults and masked by soil and vegetation, but the main outlines are still clearly visible (Fig. 217).

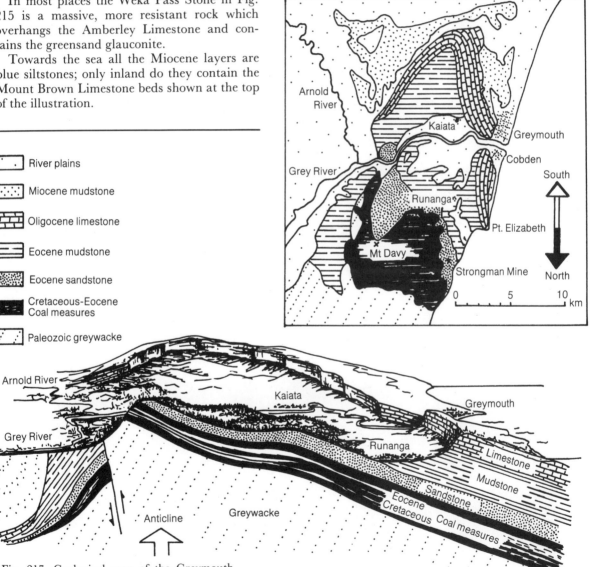

River plains

Miocene mudstone

Oligocene limestone

Eocene mudstone

Eocene sandstone

Cretaceous-Eocene Coal measures

Paleozoic greywacke

Fig. 217. Geological map of the Greymouth area (*after Bowen*), and diagram.

Fig. 218. Brunner coke ovens by the Grey River. Cretaceous-Eocene coal measures form the cliffs above and are topped by a hard layer of Eocene sandstone. (*Photograph, L. Beckford*)

Fig. 219. From the coast north of Rapahoe the Kaiata Mudstones extend south to Point Elizabeth, which is capped by Oligocene limestone. A walkway leads along the cliffs to Cobden.

Fig. 220. Cobden Limestone at Greymouth.

The underlying basement rocks are Ordovician greywackes, greenish-grey to dark grey sandstones and argillites forming the Paparoa Range north of a line from Blackball to Thirteen Mile Beach. Lapping on to the high plateau are the Cretaceous Paparoa Coal Measures, made up of layers of mudstone, sandstones and coal like those at the Strongman Mine. The basins for the coal swamps may have formed as part of the final rifting associated with the Tasman Sea opening, producing rift valleys here on the West Coast when there was little movement in the rest of the South Island. See Figs. 167 and 168.

The Paparoa Coal Measures are present only near Greymouth, but the younger Brunner Coal Measures are spread widely over the West Coast under most of the younger rocks. They are generally those layers of quartz, sand and conglomerate that were forming as the Cretaceous land weathered (page 87) and contain at least one massive coal seam such as the one first mined at Brunner, where today the Historic Places Trust is restoring the coke ovens and the rest of the mining relics as part of an industrial park (Fig. 218).

Fossil pollens in the Brunner coals are of Paleocene to Eocene age, showing that the coal swamps lasted into the time of the Bortonian Stage in the New Zealand Series (see the time listing below).

After these rocks were formed the seas moved in. On top of the Brunner layers is a thick layer of hard, marine sandstone (Island Sandstone) with plant fragments in the lower part and shell fossils higher up. This layer caps and protects some of the Brunner layers and slopes up from the softer mudstone around it.

The Kaiata Mudstone (see Fig. 217) built up as the seas deepened. It is comprised of brown mud, with some beds of conglomerate and sandstone towards the old shoreline in the east. It represents what was going on during the rest of the Eocene Epoch. The Kaiatan and Runangan Stages of the Arnold Series are based on foram sequences collected from the Kaiata Mudstone on the north side of Point Elizabeth (Fig. 219).

At Point Elizabeth the Kaiata Mudstone is overlaid by a younger rock — Cobden Limestone, a fine-grained, creamy-white to light brown, muddy limestone. Being harder than the rocks above and below it, this limestone forms the cliffs or scarps that can be traced right around the area, lifting above the eroded and slumped mudstone and forming the cliffs in the Grey River gorge behind the town (Fig. 220). The Cobden Limestone is Oligocene, the same age as the Weka Pass Stone in the Waipara Gorge.

On top of the limestone are even younger layers of soft, blue-grey mudstone, which was named the Blue Bottom by goldminers and which in places contains fossil shells. This massive mudstone is Miocene in age and borders the ranges south to the Arahura and also in patches north to Westport. In the Little Totara River it has yielded fossil cockles, tusk shells, rare nautiloids and the gross claws of the ghost shrimp (Fig. 221).

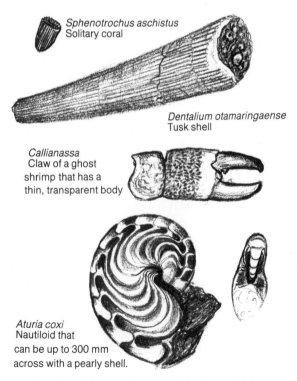

Sphenotrochus aschistus
Solitary coral

Dentalium otamaringaense
Tusk shell

Callianassa
Claw of a ghost shrimp that has a thin, transparent body

Aturia coxi
Nautiloid that can be up to 300 mm across with a pearly shell.

Fig. 221. Upper Miocene fossils from Little Totara River.

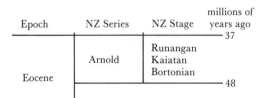

Epoch	NZ Series	NZ Stage	millions of years ago
			37
Eocene	Arnold	Runangan Kaiatan Bortonian	
			48

Fig. 222. Granite boulders on Eocene sedimentary layers, on the beach north of Cape Foulwind light.

Cape Foulwind near Westport is composed of very old Devonian granite, part of the Karamea Granite and containing the familiar pink crystals of feldspars which are often aligned. It can be seen at Tauranga Bay (see Fig. 77), at the north end by the seal colony.

Round the cape, granite boulders lie on ledges of sediments dipping down to the north (Fig. 222). The cliffs are younger as they progress north along the beach. At the south end of the beach by the waterfall there are Eocene (Runangan) corals, balls of algae and tusk shells in mudstones like the Kaiata Mudstone. Where the cliffs turn out to the reef and caves they contain rare sharks' teeth and more common corals resembling 50-millimetre-long pencils (see Fig. 223). Above the mudstones are younger sandstones, then Oligocene limestone (corresponding to the Cobden Limestone), the basis for the cement works on the terrace above. The same limestone forms Pancake Rocks at Punakaiki, where the sea has carved pinnacles and caves for our delight.

The Brunner coals are mined at Charleston in several open pits; they are visible on the roadside when travelling north. The main coalmines in this area are on the plateau north of Westport, at Denniston and Stockton (Fig. 224). There is an excellent coal museum at Westport.

On top of the coal measures, and interfingering with some of them, are the familiar Eocene mudstones of this area, although in most places on the plateau they have been eroded away from

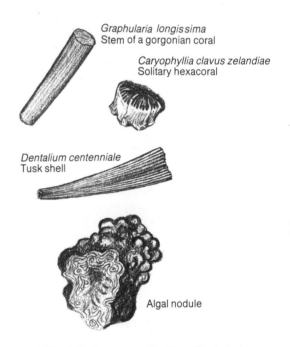

Graphularia longissima
Stem of a gorgonian coral

Caryophyllia clavus zelandiae
Solitary hexacoral

Dentalium centenniale
Tusk shell

Algal nodule

Fig. 223. Eocene fossils, Cape Foulwind.

the hard sandstone that makes up the plateau's present surface.

From the time of the Eocene Epoch the sea moved inland across the Paparoa-Buller area and into a sound further east in the Maruia-Murchison-Matiri depression. There, too, are thin coal measures, Arnold mudstones, Oli-

Fig. 224. Coal seam on Stockton plateau.

gocene limestones and Miocene marine sand-stones and siltstones.

In the ancient Murchison sound the formations are thin at the edges, but reach very great thicknesses (up to 6 kilometres) in the centre. There are variations because of local relief, but as the seas deepened they moved further on to areas where there had been no low-lying swamps.

At School Creek, south of Murchison, there is a quarry in steeply dipping Oligocene limestone, which was deposited directly onto the eroded granite sea-floor. Where the limestone butts against the granite at the south side of the quarry there is a layer of granite pebbles and debris cemented by algal mats resting on the

Fig. 225. Algae in Oligocene limestone, School Creek, near Murchison.

granite. The limestone (Fig. 225) is fine, white and contains fossil algae and small forams, typical Oligocene.

The once-level layers of sediment have been folded to a near vertical position. Travellers in the area north of Murchison, cannot fail to notice the great ribs standing up on the hills (Fig. 226). Here the Oligocene sediments are sandstone and siltstones, showing that the nearby land was then still shedding sediment. This process continued into the Miocene Epoch but by then the Paparoa and Victoria Ranges were rising, and the Grey-Inangahua depression was the main seaway. The sea left the Murchison area, and the later Miocene deposits were formed as rivers built deltas out across the old sea floor.

At the entrance to Mangles Valley, about 5 kilometres north of Murchison, and across the un-named side bridge 6 kilometres north of Longford bridge the conglomerates, sandstones and mudstones contain numerous leaf impres-

Fig. 226. Miocene sandstone ribs north of Murchison.

Cryptocarya longfordiensis

Nothofagus novozelandiae
— a beech

Fig. 227.
Upper Miocene leaves,
Longford Formation.

Fig. 228. Hematite from Parapara Inlet.

they passed over a large area of lime (the local marble) and the iron was precipitated out. There are here some 9 million tonnes of ore, which today is mined occasionally for use in the cement and gas industries from a quarry up the road opposite the road to Parapara Beach.

At Abel Head at the base of Farewell Spit we can see the effects of the later incoming of the sea in layers of coal, greensand and siltstone (Fig. 229). Whalebones have been found here, and

sions from a vanished forest (Fig. 227). They include plants that no longer grow in New Zealand, including a large-leaved beech (*Nothofagus*) which is now found only in New Caledonia.

In places the river plains which formed after the seas retreated produced thin coal seams, just like those made on the earlier lowlands.

Golden Bay — the layers on top of the oldest rocks

Further north, while the Rangitata land was being worn away during the late Cretaceous, thick conglomerates (at Wharariki Beach) and coal measures were forming at Pakawau and Puponga. The old peneplain surface south of Rockville and Parapara Inlet was spread with quartz conglomerates laced with gold — the "quartz wash" of the miners. During the Eocene the river plains persisted right across the area to Takaka, laying down widespread coal measures.

Near Onekaka the Eocene deposits include iron ore made up of the oxides goethite (which leaves a yellow crush mark on an unglazed tile or piece of quartz) and shiny, black hematite, Fig. 228 (which leaves a red crush mark). The iron deposit was formed when acidic waters containing dissolved iron were neutralised as

Opissaster sp.

Fig. 229. Freshwater layers underlie marine layers of sediment at Abel Head; Oligocene heart urchins weather out on the shore to the north.

Fig. 230. Tarakohe Cement Works quarry, showing grey siltstone to the left above the stripped surface of the Oligocene limestone.

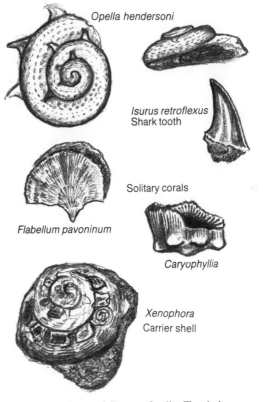

Opella hendersoni

Isurus retroflexus
Shark tooth

Flabellum pavoninum

Solitary corals

Caryophyllia

Xenophora
Carrier shell

Fig. 231. Lower Miocene fossils, Tarakohe.

little heart urchins weather out of concretions in a grey siltstone. These rocks are Oligocene. By the end of this epoch the seas were depositing a layered limestone containing many shell fragments. It mantles much of Golden Bay and is eroded into caves at Rockville which are open to the public.

Late Oligocene limestone is the main raw material for the cement works at Tarakohe, just as the limestone of the same age is used at Cape Foulwind. It is covered by a younger, soft, grey siltstone, laid down as the land buckled up in the Miocene. While the Mangles area was rising, the Golden Bay area deepened and filled with the extra silts being brought down from the new uplands.

In the quarry (Fig. 230) the siltstone has been stripped off the surface of the limestone. The piles of dumped material have yielded many Miocene fossils (Fig. 231), including the only large fossilised crayfish so far found in New Zealand.

Grey siltstones also appear in cuttings on the road to Collingwood, 6–10 kilometres north-west of Takaka.

In the main Takaka Valley (Fig. 232) the cover layer of younger rocks has been gently folded and eroded, exposing the Eocene coal

Miocene mudstone

Oligocene limestone

Eocene coal measures

Fig. 232. Tertiary sediments gently folded over the old basement in the lower Takaka Valley.

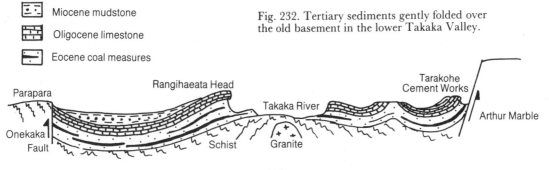

Parapara

Rangihaeata Head

Takaka River

Tarakohe
Cement Works

Arthur Marble

Onekaka
Fault

Schist

Granite

Fig. 233. Coal seams at the base of the bank in the foreground and limestone forming the cliffs of Rangihaeata Head beyond. (Look for quartz with green chromium mica on the beach between.)

measures near Pupu Springs and down by the beach east of Rangihaeata Head at the mouth of the Takaka River (Fig. 233). Here coal seams are present in the cliffs at sea level, and as you walk along the beach west to the bluff you can see how the layers gradually change through delta and river gravels and sands to Oligocene marine limestones on the head, while to the west of the head the shore platform reveals gently dipping layers of Miocene mudstones.

The Landon Series — Oligocene
North Otago

Epoch	NZ Series	NZ Stage	Millions of years ago
			— 22.5
Oligocene	Landon	Waitakian Duntroonian Whaingaroan	
			— 37

So far we have looked at Oligocene limestones at Waipara, on the West Coast and in Golden Bay. They occur all over the country, but the area inland from Oamaru is the type section, although the lowest stage, the Whaingaroan, is named from Whaingaroa Harbour, Raglan, as it is typical of the North Island deposits of this time.

As usual, we will look at the older rocks in the area first.

North of Palmerston the basement schists have been covered by Cretaceous river sediments with small, workable coal seams at Shag

Point. We know that even in Cretaceous times the seas were moving in because the first thick, blue-grey siltstones and mudstones near Katiki are also Cretaceous. Some concretions in these siltstones and mudstones contain reptile fossils, including the plesiosaur (Fig. 234) recently excavated by Otago University geologists.

Further north, the magnificent concretions known as the Moeraki Boulders (Plate 4C) were formed in the Paleocene, as shown by the forams in the mudstone surrounding them; the muddy sea-floor conditions persisted from the Cretaceous to the Eocene. Golden calcite crystals are visible in the septarian cores.

Fig. 234. Concretion encasing a plesiosaur lying on the beach at Shag Point before its removal to the University of Otago. The head was buried in the sandstone bank to the right, and the rounded mass encased the body.

During the Eocene the first outpourings from basaltic volcanoes went into the sea, creating pillow lavas shaped like those at Red Rocks in Wellington. At Oamaru they are much fresher and more accessible than at Red Rocks. At Boatmans Harbour, a 10-minute walk from the breakwater at the south-east end of the main harbour, the beach is backed by a wall of pillow lavas (Plate 6G) with basalt cores and glassy, greenish rims. When the lavas oozed into the shell debris on the sea floor, scraps of limestone were caught between the pillows (Fig. 236). On the northern end of the beach the folded layers of volcanic ashes also contain fossil shells.

As soon as the eruptions died down, the piles of volcanic debris formed a platform where shellfish flourished — molluscs and lampshells sea urchins and bryozoans, the lacy moss-animals. Their shells built up to make pure, creamy limestone which has been widely used as a building stone since the 1860s, turning Oamaru into a white stone city. Today Oamaru stone comes from Parkside Quarry (Fig. 237) near Weston, also in the same Eocene limestone formation.

Mineral hunters are especially rewarded by the volcanics in this area because a second series of explosive eruptions produced a coarse breccia, which in the cuttings and roadsides at Oamaru Harbour and at the quarry opposite the old water wheel contains many cavities with tiny calcite and zeolite crystals. On the beaches south and north of Kakanui the beach rocks are

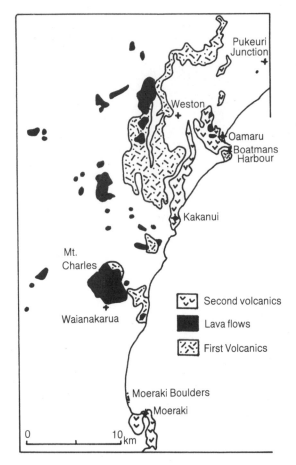

Fig. 235. Map of volcanics near Oamaru (*after Mutch*).

Fig. 236. Detail of pillow lavas at Boatmans Harbour, showing glassy rim and limestone between the pillows.

Fig. 237. Oamaru stone at Parkside Quarry, Weston.

Fig. 238. Kakanui Breccia in close-up, with a central nodule of green olivine.

Fig. 239. Xenoliths from the Kakanui Breccia. Top, Rounded nodule of kaersutite, black titanium-hornblende, and broken crystal showing cleavage. Bottom, colourless anorthoclase feldspar — a rounded nodule showing broken end with cleavage and some attached breccia.

Kakanui Breccia containing crystals of garnet, red and green spinel, white anorthoclase feldspar, shining kaersutite (hornblende) with good cleavage and black, high temperature augite without obvious cleavage (Fig. 239). Nodules of olivine in the breccia here are usually reddish, from weathering, rather than olive-green.

The bright-green crystals present are chrome diopside, another member of the pyroxene group.

All these minerals are contained in a weathered groundmass of ash and lava. They are foreign to the main mass of the volcanics and are therefore known as xenoliths or "stranger stones". They are thought to have come from far down in the mantle where the crystals formed in the magma over a long time while the pressure built up for a great eruption which pushed the mantle minerals up with it. Sometimes the minerals held their mantle temperatures of 1100°C or more and cooked the surrounding rock, producing a darker rim like that surrounding the olivine shown in Fig. 238. Some of the crystals and nodules were worn smooth as they were hurled up the vent. With care a few samples can be picked out of the decomposing rock, but this is a study area which should not be cleaned out.

The southernmost volcanics at Moeraki Point contain seams of agate which weather out and can be found further north on the beaches, especially on the coast past Waianakarua River.

While there take a look at the Waianakarua north bridge on State Highway 1; it was built of limestone in 1874.

The second volcanics at Oamaru are early Oligocene in age. The whole New Zealand series of Oligocene rocks was first described from Landon Creek (Fig. 240), which winds down from low hills just south of Pukeuri Junction. Today this area is overgrown with pasture and the local rifle club has its range along the type section. An outstanding display of Landon or Oligocene rocks is up the Waitaki Valley from Bortons past Duntroon, as the stage names imply (see the chart on page 120). See Fig. 241.

From the old Bortons Railway Siding the

Fig. 240. Limestone in Landon Creek.

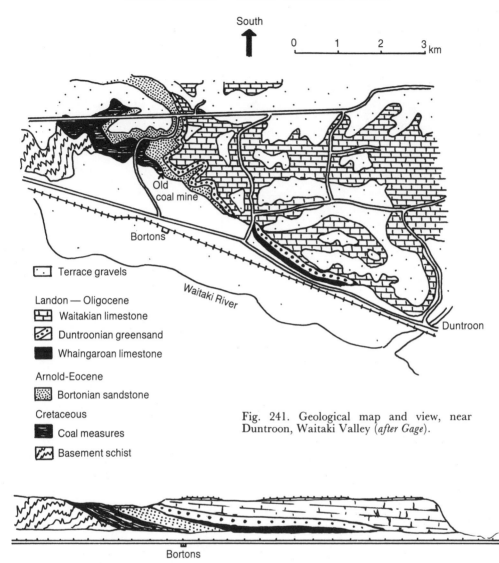

Fig. 241. Geological map and view, near Duntroon, Waitaki Valley (*after Gage*).

whole Tertiary sequence can be seen. Down the valley the cliffs are of basement schist; at Bortons a side road passes thin Cretaceous coal measures with an old coalmine. On top of these are the thick Eocene muddy greensands which gave their name to the Bortonian Stage of the series (see table, page 115). They slump away and are visible not as rocks but as well-grassed slopes. To the west they are covered by the fine, white Oligocene limestone left by the shallow sea that had spread all over the flanks of New Zealand.

After the submergence there was little land left and only a small amount of land-derived sediment reached the sea, so the sea floor was mostly an ooze of fine shells and shell pieces, forams and bryozoans, algae, corals, crustaceans and sea urchins, molluscs and brachiopods with glauconite greensand forming among them (Fig. 242). See Plate 5A.

In the softer greensand layers, as in the Waipara cliffs, there are fossil bones, the earliest remains found to date of whales, dolphins, penguins and other sea birds. The penguins of this time were giants more than a metre tall. They are known from a few bones only; most were collected in the low cliffs of the Waitaki Valley.

123

PHASE III: A PLACE OF OUR OWN

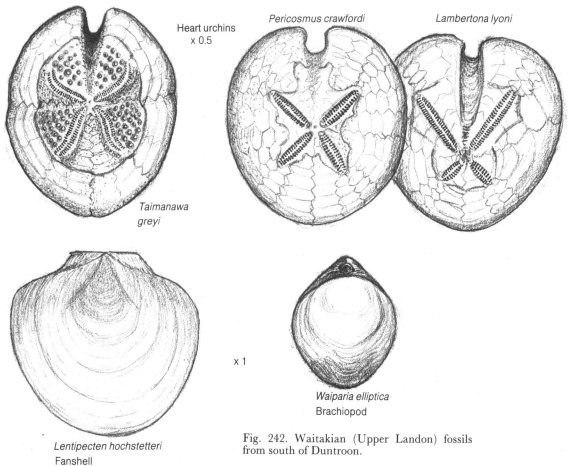

Heart urchins
x 0.5

Pericosmus crawfordi

Lambertona lyoni

Taimanawa greyi

x 1

Lentipecten hochstetteri
Fanshell

Waiparia elliptica
Brachiopod

Fig. 242. Waitakian (Upper Landon) fossils from south of Duntroon.

Fig. 243. Limestone bluff with Maori cave drawings west of Duntroon.

The white limestones laid down in Landon times have hardly been tilted at all; they form flat-topped bluffs all along the valley and inland from Duntroon (Fig. 243). They have been used as building stone and as fertiliser, and on their white, curved surfaces the early Maori painted with red earth and charcoal.

At some places the limestone is sandy, and at Waihao Forks and Waihao Downs (Fig. 244)

great cross-beds are visible in the cliffs where currents scoured out troughs on the sea floor which were filled as the sand and shell banks shifted sideways.

These white limestones built up in the shallow seas stretching most of the length of the east coast, eventually covering much of the former land.

Fig. 244. Above, cross-beds in the limestone at Waihao Downs. Below, limestone and kiln by Pareora River.

LATE TERTIARY — GOING GREY

During the Miocene Epoch in most areas the white Oligocene limestones were covered with grey siltstones and sandstones.

The Pareora Series — Early Miocene

South Canterbury

In South Canterbury many of the farm houses nestle against limestone bluffs which yielded the creamy stone that was cut for their building. Near Pareora, as at Kakahu, a lime kiln was built to burn lime for fertiliser.

From the Lower Pareora Gorge the limestone dips gently south-east, covering the Eocene rock in the gorge, as shown in the cross-section illustrated in Fig. 245. On top of the limestone is a younger, blue-grey siltstone with some different fossils forming the low, grass-covered hills south of the cluster of fishermen's cottages which lie about 6 kilometres west of Southburn. This is covered in turn by Altonian layers of creamy sandstone with hard bands which form a bluff downstream (Fig. 246). The sequence of rocks is most clearly shown here by the Pareora River, but the Otaian siltstones by the fishermen's cottages are too slumped to form a type section.

If you are in this area, while you are here look north across the river to the cliffs of Mount Horrible. These were formed by the last lava flows in the South Island, which, during the Pleistocene Epoch, poured down to the sea at Timaru. The basalt layer has protected the softer Pareora Series rocks underneath.

The type section for the Otaian Stage is found where the Pareora siltstones and sandstones are much more exposed, a few kilometres to the south in the Otaio River at Bluecliffs (Fig. 247), about 2 kilometres upstream from Drinnans Bridge or a kilometre downstream from the ford at McAlwees Crossing. It is probably easier to walk downstream than across the farmland because there is a belt of tall gorse along the river bank.

Several species of shellfish have evolved and been found in the blue-grey siltstones here (Fig. 248), and it is the fossils rather than the change of rock type which determine the Pareora Series, although this blue-grey mudstone and siltstone is typical of Miocene marine sediments around New Zealand.

The Pareora Series once contained a further two stages, the Awamoan and Hutchinsonian, which now are known to have been based on groups of animals that developed not because of the passage of time but because of special local conditions. Some of these fossils were found in the Waitaki Valley, in an outcrop in the river

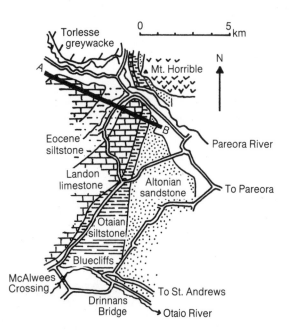

Fig. 245. Cross-section and map for Lower Miocene Pareora Series, Pareora and Otaio Rivers (*after Gair*).

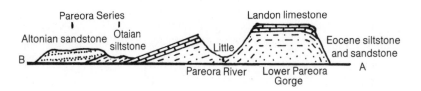

	NZ Series	NZ Stage	millions of years ago
			─17
Miocene	Pareora	Altonian	
		Otaian	
			─22.5

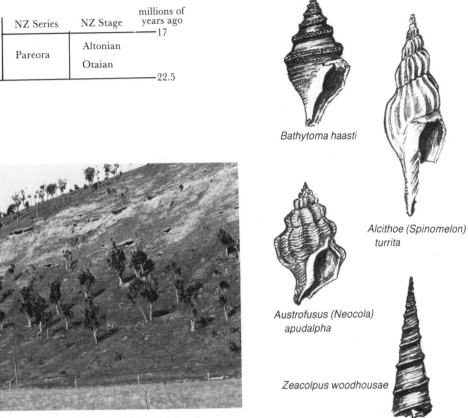

Bathytoma haasti

Alcithoe (Spinomelon)
turrita

Austrofusus (Neocola)
apudalpha

Zeacolpus woodhousae

Fig. 246. Altonian sandstone with harder bands beside the Pareora River.

Fig. 248. Otaian fossils from the Bluecliffs Silt, Bluecliffs.

Fig. 247. Bluecliffs siltstone, Otaio River.

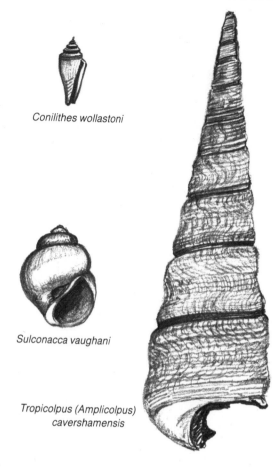

Conilithes wollastoni

Sulconacca vaughani

*Tropicolpus (Amplicolpus)
caivershamensis*

Fig. 249. Altonian fossils from the Waitaki River bank, Peebles.

The Southland Series — Middle Miocene

Southland

The coal-producing swamps that existed at Ohai at the end of the Cretaceous and in the Eocene have already been mentioned. In ancient Southland a pattern of land and sea may have been created by movement on old faults which lifted up Fiordland and the Takitimu Mountains to make a deep basin between (Fig. 250), edged perhaps by the same kind of fault scarp as that along the Hutt Road in Wellington.

At first this deep, western valley was above sea level, so that while the coal seams were forming at Ohai in the Eocene, rivers and flood plains of the western basin were filling with thick, freshwater deposits — gravels, sands and silts — but only thin coal seams were formed.

This basin, stretching from the Waiau Valley north to the Eglinton Valley, lasted right through the Tertiary Period. By the Oligocene Epoch the sea had moved in, and the thick layers of mud and sand coming down from the surrounding hills supported marine organisms. They were mostly minute, but in several places there were banks of shellfish and shell debris which led to the formation of limestones in those areas. The hills above the Eglinton Valley are capped with these marine layers.

bed itself about 1 kilometre east of the end of Jardine Road, north of Peebles (Fig. 249). These animals were not typical of distinct periods of time, and the stages formerly based on them have now been replaced by one stage from the middle Miocene Southland Series, which is marked out in the rocks of Southland where there was a widespread sea at this time.

Fig. 250. Southland in the early Tertiary Period.

Fig. 251. Oligocene limestone hill, Bobs Cove, Lake Wakatipu.

The Te Anau-au Caves on the west of Lake Te Anau below the Murchison Mountains are in Oligocene limestones; these limestones also form the white rock that lies on the roadside and fields near Whitestone River, between Te Anau and Mossburn.

A second long embayment reached even further north and can be seen today in the sandstone-limestone hill at Bobs Cove (Fig. 251) on Lake Wakatipu, about 15 kilometres west of Queenstown on the road to Glenorchy. The Oligocene sediments of this long sound have been preserved all along Moonlight Fault, a weakness which runs from the Oreti to the Shotover Rivers. At Bobs Cove along the lakeshore to the east of the hill the lowest Tertiary sediments are breccias which rest on schist and contain schist pebbles coated with algae, algal nodules and large forams. Above

these are layers of sandstone containing shallow-water shellfish fossils, and some thick beds of limestone.

In eastern Southland the shallow Oligocene-to-Miocene seas stretched away past the site of Winton, but in the Hedgehope and Mataura areas there were low hills and a plain where rivers meandered, spreading layers of silts, gravels and plant debris which made up the Gore Lignite Measures. Today these are being explored for future development on a wide scale; they have been mined at Pomahaka and Waimumu for local use for many years.

The first low hills north of Pukerau (Fig. 252), just east of Gore, are capped with white quartzite boulders; looking east past them we can see the Permian valley towards Arthurton and the Triassic ridges beyond, jutting against the sky. In the early Miocene this area was

Fig. 252. Quartzite boulders on the hill north of Pukerau.

Casuarina cone impression

Tainui leaf

Fig. 253. Plant fossils, Pukerau.

Where the rivers carried less sand the sea floor was covered with shell debris causing limestones to build up, but this happened later in Southland than elsewhere. The limestones that provide fertilisers for the lush Southland plains are mostly early Miocene or Pareoran in age. At Castle Rock (Fig. 254), Browns, Winton, Forest Hill, Limehills and Clifden the creamy limestone in the quarries contains mostly a fairly coarse shell debris with occasional whole shells, pectens, brachiopods (Fig. 255) and rarer sharks' teeth.

covered with white quartz sands washed off the schist uplands, and the rivers that left the sands also carried down plant debris — branches, leaves and seeds — from the surrounding low hills and swamps. In time the sands were cemented by groundwaters rich in silica, and the plants were preserved as casts or impressions with a few rare, silicified parts (Fig. 253). They are of great value to those who study ancient plants (paleobotanists) as they are not squashed flat; rubber casts taken from the quartzite moulds show the three-dimensional shapes of the seeds.

The plants include some ferns, kauri, rimu, beech, tainui, lacebark and many others. The Australian she-oak, *Casuarina*, had an ancestor growing here, too, as moulds of its cones have been found, indicating a warmer climate than that of today.

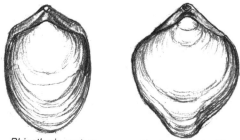

Rhizothyris scutum   *Pachymagus parki or Waiparia*

Fig. 255. Brachiopods from Forest Hill, Southland.

At Clifden the banks of the Waiau above the historic swing bridge contain a continuous series of layers on top of the limestones. The fossils in these rocks enabled geologists to establish ages for the different layers, which are the type sections for the middle Miocene (Fig. 256).

	NZ Series	NZ Stage	millions of years ago
Miocene	Southland	Waiauan Lillburnian Clifdenian	— 12 — 17

Fig. 254. Limestone near Castle Rock, Southland.

Access to the banks is tricky as the Waiau flows fast and deep at the base of some of the cliffs (Fig. 257). The geologists studying the section cut paths down through the bush to the beaches; one even swung on rope ladders to collect specimens from some layers of shells.

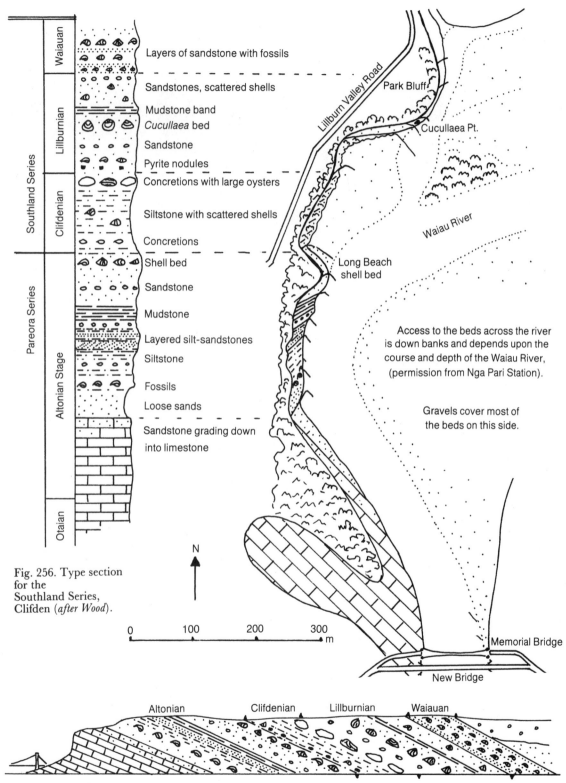

Waiauan — Layers of sandstone with fossils

Lillburnian — Sandstones, scattered shells

Mudstone band

Cucullaea bed

Sandstone

Pyrite nodules

Clifdenian — Concretions with large oysters

Siltstone with scattered shells

Concretions

Shell bed

Sandstone

Mudstone

Layered silt-sandstones

Siltstone

Fossils

Loose sands

Sandstone grading down into limestone

Southland Series

Pareora Series

Altonian Stage

Otaian

Lillburn Valley Road

Park Bluff

Cucullaea Pt.

Waiau River

Long Beach shell bed

Access to the beds across the river is down banks and depends upon the course and depth of the Waiau River, (permission from Nga Pari Station).

Gravels cover most of the beds on this side.

N

Fig. 256. Type section for the Southland Series, Clifden (*after Wood*).

0   100   200   300
                          m

Memorial Bridge

New Bridge

Altonian   Clifdenian   Lillburnian   Waiauan

131

Fig. 257. Southland Series, looking upstream from Clifden Memorial Bridge.

The lower and older rocks are sandy limestones, with less sand and more animal remains in the Clifden Limestone, which contains corals, bryozoans and shellfish (Fig. 258). On top of this lie the rocks of the rest of the Southland Series — all grey sandstones, siltstones and mudstones with some shells and concretions.

Such beds are being laid down on the coastal sea floor today. Nearly all the Miocene marine deposits of New Zealand are similar blue-grey sandstone, siltstone and mudstone, showing that rivers were once again carrying debris to the sea.

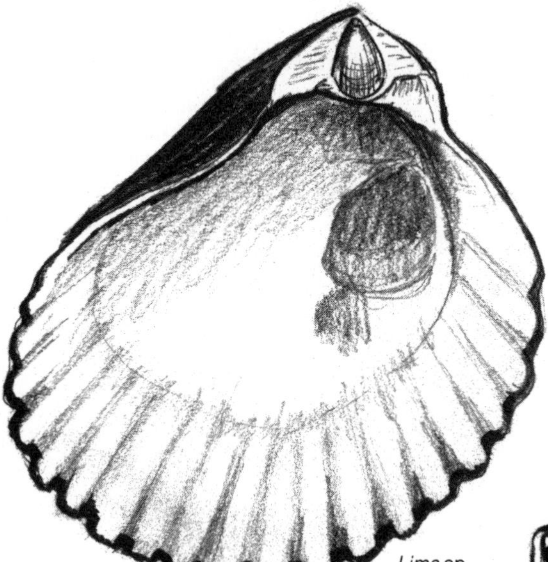

Lima sp.

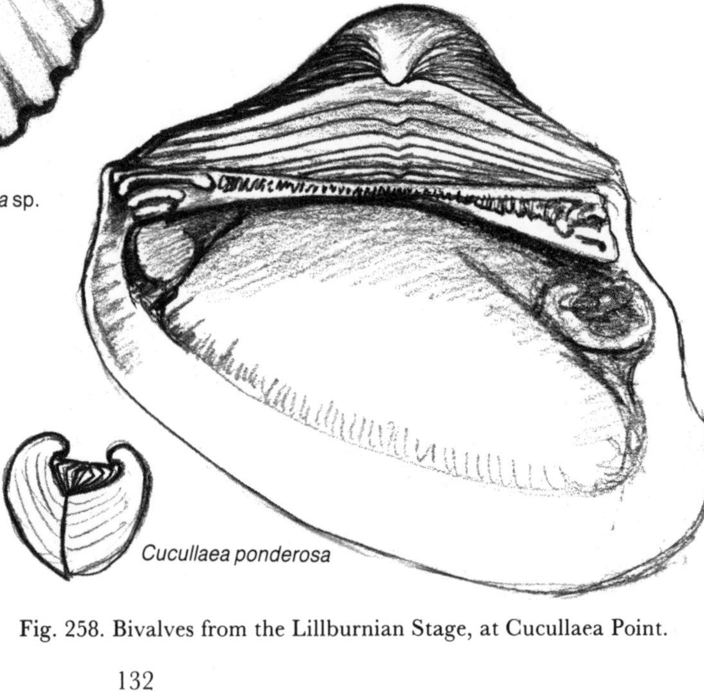

Cucullaea ponderosa

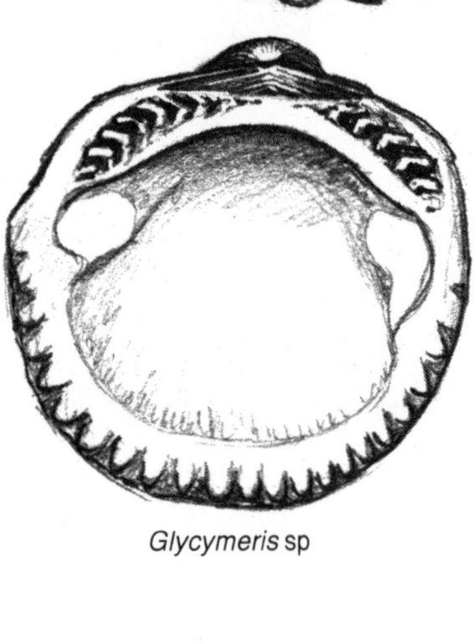

Glycymeris sp

Fig. 258. Bivalves from the Lillburnian Stage, at Cucullaea Point.

Central Otago — the Miocene landmass

While the Southland area was covered by sea, like much of the rest of the land at the time, Central Otago was a gently rolling land of schist hills drained by rivers flowing into inland lakes before making their final way to the sea. The rivers dumped their debris at the edges of the lakes to build deltas. In shallow backwaters the freshwater mussel thrived, and on the flood plains were peat-forming swamps (Fig. 259).

The silts, muds and coals (known as the Manuherikia Group of rocks, see Fig. 262) are present in small pockets all over Central Otago, preserved in fault-angle depressions high on the mountains as well as in the valleys.

When a covering layer is pushed down against an uplifted block, erosion on the higher ground covers and protects the side that has been pushed down (Fig. 260) so that when the entire top layer is eventually removed the small, lower part is left as a clue to what may once have been a widespread deposit.

Fig. 261. Bannockburn.

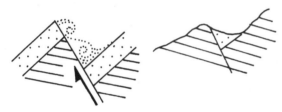

Fig. 260. Preservation of sediments in a fault-angle depression.

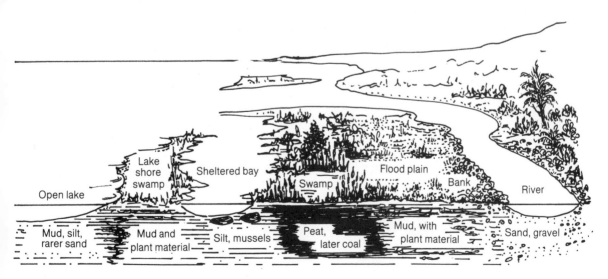

Fig. 259. Central Otago in the Miocene (*modified after Douglas*).

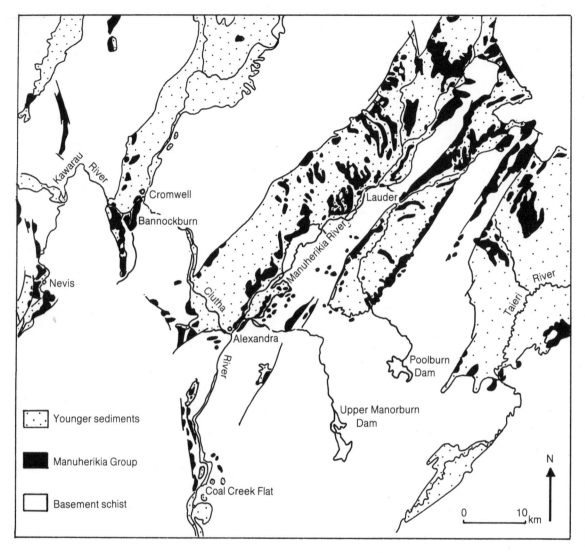

Fig. 262. Localities of the Manuherikia Group of freshwater sediments (*after Wood and Mutch*).

There are slivers of coal measures high above Gibbston in the Kawarau Gorge and at Doolan Saddle in the Nevis. There are coal pits near Wedderburn, and at Coal Creek by Roxburgh Dam in Harliwich's Pit there is the thickest coal seam in the country, measuring 45 metres.

In the Nevis Valley the shales contain oil, although at present it would take more energy to heat the shales and release the oil than the oil is worth (Plate 5E).

Down Cairnmuir Road near Bannockburn (Fig. 261) collectors search for gypsum crystals in the dark mudstone and split the shales for fossil leaves from the plants that grew around the swamps and in the nearby hills (Fig. 263). Some of the best impressions of *Casuarina* cones have been found here. Recently, in similar shales near St Bathans, a beetle larva was found.

The lowest layers of sediments here include white quartz sands and gravels which built up during the long weathering of the Cretaceous landmass. In places groundwaters have cemented these into conglomerates and quartzite like the Pukerau stones but without plant fossils. These boulders now lie on the surface of the schist and mark where a covering of softer

Fig. 264. Ventifact near Poolburn, Ida Valley.

Fig. 263. Bannockburn gypsum, *Casuarina* cone and leaves and ancestal lacebark leaf.

Tertiary sediments has been eroded away.

Although they are very hard and resistant to water erosion, the strong winds of the last Ice Age have sandblasted the exposed rocks, shaping them into the characteristic ridged form of a ventifact (Fig. 264), meaning "wind made", and giving them surface lustre.

Beyond Naseby in the northern part of this land area the hills lifted by the Rangitata Orogeny remained a barrier to the sea. It has already been mentioned (page 81) that the breccias at Kyeburn were part of the evidence for a great uplift at the beginning of the Cretaceous Period; the Kyeburn Breccia forms the low, rounded hill to the west of the Kyeburn River (Fig. 265) and, when seen from the south, echoes in its layers the slope of the eroded surface of the Ida Range behind. The mountains to the right, below Mount Kyeburn, are of schist. A prominent ridge formed by harder

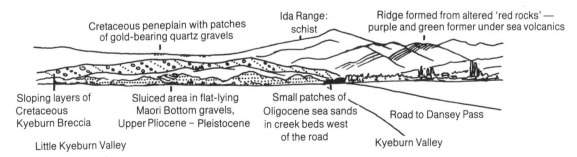

Cretaceous peneplain with patches of gold-bearing quartz gravels

Ida Range: schist

Ridge formed from altered 'red rocks' — purple and green former under sea volcanics

Sloping layers of Cretaceous Kyeburn Breccia

Sluiced area in flat-lying Maori Bottom gravels, Upper Pliocene – Pleistocene

Small patches of Oligocene sea sands in creek beds west of the road

Road to Dansey Pass

Little Kyeburn Valley

Kyeburn Valley

Fig. 265. Geology between the Little Kyeburn and Kyeburn valleys, looking north to Dansey Pass.

Fig. 266. Detail of the schist in the river above the Dansey Pass Hotel, showing the remains of graded bedding; pale quartz-schist grading into darker schist from mudstone.

layers of altered red rocks — volcanics which provide dark-red jaspers with epidote seams in the Kyeburn River boulders — runs north and east for many kilometres. The band shows up well in the ridges up towards Dansey Pass. The schists here are unusual in that they show remains of the graded bedding in the original greywacke sediments; this is positioned at right-angles to the cleavage in the schist. These bedding layers are visible in the rocks on the banks of the stream between Dansey Pass Hotel and the German Creek bridge upstream (Fig. 266 and Plate 2A). Above them on the terraces

are the remains of gold workings.

The Kyeburn Breccia forming the cliffs on the east banks of the Little Kyeburn Stream are steeply tilted layers of gravel. They are coarser to the south-west, and they contain layers of finer silts and even peats to the north, showing that they formed a scree fan below steep cliffs (now eroded) to the south. They are known to be Cretaceous in age from pollens in the peats.

The sea that was covering the country in the Oligocene Period moved inland just as far as Naseby, where greensands only 15 metres thick contain marine shells. The same fault-angle system that gave us the scattering of coal measures in Central Otago preserved the familiar layers of coal measures, greensands and limestones along the line of the Waihemo Fault, which uplifted the Horse Range and Kakanui Mountains. The Pigroot road (Highway 85) to Central Otago from Palmerston follows this fault line; small patches of marine sands and limestone can be seen on the hills and in the roadcuts.

In the Kyeburn valley, about 1 kilometre below Dansey Pass Hotel and south of the quartz sand cliff, there is a creek on the west of the road which shows the greensands and fossils with some gypsum crystals. A fault has pushed the Kyeburn Breccia up against them and it now towers over the greensands as it erodes into pinnacles (Fig. 267).

In much of this area the greensands are covered by the same river silts and gravels and

Fig. 267. Pinnacles in the Kyeburn Breccia.

136

Fig. 268. Roadcut near Swinburn Stream: pale river silts to left; ash layers with "petrified hailstones" in centre; and basalt lava flow on the right beside the group.

thin lignite seams that started the whole layer cake on the old Cretaceous surface, showing that the land rose again from the sea.

Later, during the Pleistocene, the valleys were filled with thick layers of gravels which are not as tilted as the Kyeburn Breccia.

Near Swinburn Stream, about 600 metres east of Prestons Road and 5 kilometres east of Kyeburn settlement on The Pigroot, the road cuts through white river silts and gravels overlain (going west) by a layer of volcanic ash which is covered in turn by a basaltic lava flow (Fig. 268). The ash contains small (6-millimetre), round concretions where the exploding column above the volcano caused a thunderstorm in which hailstones built up out of layers of ash and ice. These "petrified hailstones" are called chalazoidites. They are not easy to see and will crumble when they are removed from the bank. They occur in some ash layers in the North Island as well.

The same basalts cap several hills beside The Pigroot. Large flows at Waipiata (Plate 6F) and Kokonga have been quarried; stone from Kokonga was used in the building of the Dunedin Railway Station. These volcanoes erupted about 15 million years ago, at the end of Southland time.

The Te Kuiti Group

North Island

Meanwhile in the North Island there were two contrasting areas. While the east of the island had been a sea from Cretaceous time, the western side from North Auckland to Taranaki had been going through the same events as the South Island — the land uplifted by the Rangitata Orogeny was being worn away and coal swamps developed on the lowlands until these in turn were covered by the sea. To begin with the seas deposited silts and sands brought down from the eroding land, but as the land was reduced to a few gentle hills, erosion slowed and the sea was floored with debris from the creatures that flourished in the warm water.

All the western rocks of this time, the Eocene-Oligocene (Arnold-Landon) rocks are typified by one North Island group — the Te Kuiti Group (Fig. 269). The Eocene Waikato Coal Measures are the lowest and oldest part of this group.

The oldest peat swamps formed in the early Eocene in a depression near Rotowaro, southwest of Huntly. The area must have been like the peat swamps near Hamilton today — little swampy areas amid low hills — for the coal seams made by these swamps have irregular bases and thin out from area to area. There are sands, clays and silts between the coal seams, which were deposited while the swamps were covered by flood debris.

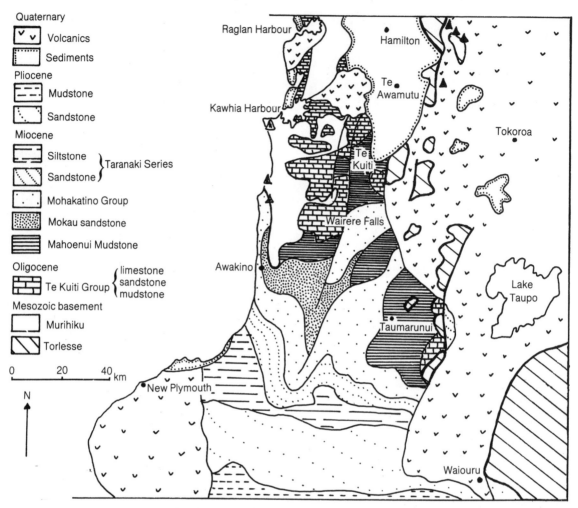

Fig. 269. Simplified geology of the King Country (*after Nelson*).

By the end of the Eocene these small patches had been merged with much wider swamp deposits, over an area from Mangapehi near Te Kuiti to Maramarua in the north and Kawhia in the west.

In the open pit at Kimihia near Huntly are the same kind of iron-red nodules as at Ohai. Here, too, they contain impressions of leaves; they also may contain seams of yellow calcite crystals in spectacular contrast to the iron-red rock. At Maramarua the calcite often takes on a "cone-in-cone" form (Fig. 270), perhaps the result of pressure during formation. It looks like a series of nested cones, but it actually has the shape of a spiral.

Fig. 270. Cone-in-cone calcite.

The Waikato coal seams are covered by siltstones and sandstones, which were deposited as the sea moved onto the land. These layers are thickest on the coast west of Hamilton, in the areas for which they are named.

The lowest or oldest siltstone was described at Whaingaroa, the first name for Raglan Harbour, where it contains a good series of microscopic fossils and has been taken as the type section for the lowest Landon or Oligocene stage. (Remember, the Landon Series up the Waitaki Valley has the Whaingaroan Stage at its base.) The pale-grey Whaingaroa Siltstone (Fig. 271) generally forms grassy slopes, and at the back of the harbour there are only a few banks on the road where it can be seen.

Above the siltstone is the Aotea Sandstone, which forms prominent bluffs on the Raglan-Kawhia road around the back of Aotea Harbour. Both the sandstone and siltstone can weather like marble as they may contain up to 50 per cent lime. A better place to look at these rocks is around Kawhia Harbour, where the Jurassic rocks are covered by the Te Kuiti Group.

Down Okoko Road, between Ngutunui and Oparau on the main road to Kawhia from Te Awamutu, the Pirongia Mine of Glen Afton Collieries provides coal for the Te Awamutu dairy factory. The single, thick seam is typical of the coal seams found in western King Country, being high in sulphur. At the top it contains hard stones and chunks of resin, and over it lies the Whaingaroa Siltstone, which here contains more shells than usual and rare nautiloids and crustaceans.

At Kawhia Harbour, west of the Puti Bridge, going south along the shore to Motutarakatua Point, the dark grey Jurassic siltstones dipping down to the shore are followed by lighter grey Whaingaroa Siltstones. Above, and on the shore further along, is the Aotea Sandstone with its widespread worm borings and tracks and layers containing fanshells and sometimes heart urchins (Fig. 272).

The layers of the Te Kuiti Group vary from place to place, just as today the seashore varies round the coast. In the northern Waikato and in the east siltstones continued to build up, but in the south and west patches of limestone and greensands were formed. Gradually the shell beds spread until the whole area was covered with a thick, creamy limestone — Otorohanga

Fig. 271. Pale grey Whaingaroa Siltstone, Ranui, Raglan Harbour.

Fig. 272. Worm borings and fanshells in Aotea Sandstone, Motutarakatua Point, Kawhia.

Fig. 273.
Totoro Gorge.

Limestone, the equivalent of the Landon limestones in the South Island.

There are many caves and systems eroded in this limestone, the best-known being at Waitomo where there is an excellent museum about caves. Near Te Kuiti the limestone is quarried for industrial and agricultural use.

Around Te Kuiti itself and to the south the landscape is covered with rounded hills of younger mudstone, the Pareora-age Mahoenui Mudstone (see Fig. 269). This slumps badly because of its high clay content. Where the mudstones are present in just a thin layer on top of limestones, as in the Totoro Gorge south-west of Aria, the limestones form sculptured shapes on the lower slopes of the hills (Fig. 273).

Going south through the King Country the rocks are progressively younger, although the sequence is repeated because faults have raised the older rocks back to the surface. South from Piopio on the Te Kuiti-New Plymouth State Highway 3 is more limestone country, especially in the Mangaotaki Gorge where the cliffs show all the rocks of the Te Kuiti Group.

On the side road past Wairere Falls the limestones can be seen sitting directly on top of the Triassic rock that forms the basement here; just past the falls the entire local sequence is visible in a quarry with Triassic greywacke at the bottom, Oligocene limestone in the middle and the Miocene Mahoenui Mudstone on top. Since the lower layers of the Te Kuiti Group are missing here, this area must have been high land while the coal seams, siltstones and sandstones were being laid down to the north.

To the west of Wairere Falls, along Ngatama-

Fig. 274. Aragonite crystals, serpentine quarry, Ngatamahine Road.

Fig. 275.
Heart urchin,
Te Maire.

Fig. 276.
Lower Miocene cliffs,
Raurimu.

hine Road, there is a large serpentine and limestone quarry (Plate 5D). It is thought that this is part of the old ultramafic belt in the deeply-buried junction between the Torlesse and the Murihiku rocks which is here squeezed up into younger limestones. To the west of the quarry the limestone is steeply dipping, but on the east it lies level and contains pebbles of serpentine, so when the serpentine came to the surface is uncertain. Greenish-white aragonite crystals (Fig. 274) have been found here, and the east face of the quarry shows a real layer-cake of rocks.

To the south-east, near Taumarunui, the Mahoenui sediments change character. These lower Miocene rocks form great cliffs of banded siltstones and mudstones showing graded bedding and some plant fossil fragments. They are unstable cliffs with few fossils although, near Te Maire, poor, flattened heart urchin and crab fossils can be found (Fig. 275). The photograph (Fig. 276) shows the alternating layers of siltstone and mudstone near Raurimu.

Back on State Highway 3, south of Mahoenui, the Pareora mudstones dip under Mokau Sandstone, of Southland Series age. This muddy sandstone contains some bands of conglomerate and it makes a hard cap over the softer mudstones underneath and forms prominent cliffs.

All the rocks we have been looking at in this section are close together at the Awakino Tunnel (Fig. 277). Stopping east of the tunnel

you can see the Mokau Sandstone cliffs to the south above the Mahoenui Mudstone slopes. The tunnel is cut in the Te Kuiti Group limestone which dips steeply east; across the river to the north is a bluff of the same rock showing its true slope.

On the western end of the tunnel you can see the details in the limestone (Fig. 278). Only the top 7 metres is Otorohanga Limestone, or the main limestone of the area — a fine-grained bryozoan material. Below that is a 38-metre layer of an older limestone containing some sandy layers and large shell fragments and conglomerates. This is Orahiri Limestone and is sometimes differentiated from the Otorohanga Limestone by the presence of large oyster fossils.

Both limestones are flaggy, or splitting apart into layers. These are not the more common layers made by differences in the sediments but the result of the rock-forming process when the sediments hardened and were cemented together. As layers of sediment built up pressure upon the limy shell fragments, the shells tended to dissolve on pressure points and the dissolved calcite was carried away by groundwaters to layers that were better supported, leaving weakened areas which later split.

Here the splits run in two directions. Those formed first are parallel to the original layers in the limestone, which has since been tilted and overlaid with sandstones. Then a second set of splits formed parallel to the layers of Mokau sandstones, showing that these rocks also ex-

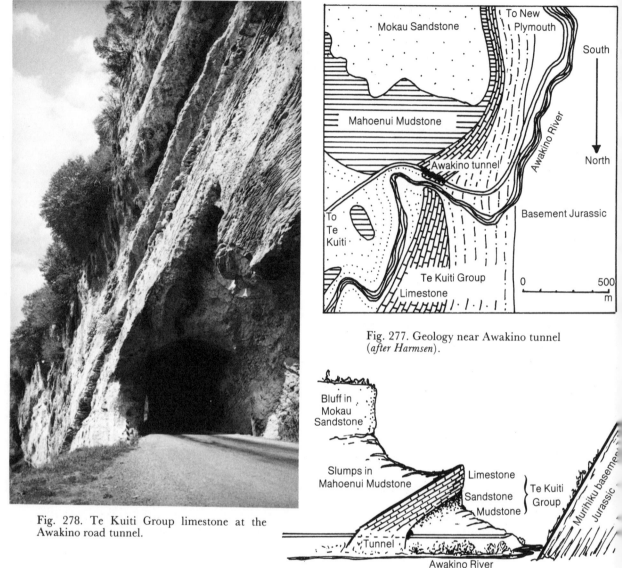

Fig. 277. Geology near Awakino tunnel (*after Harmsen*).

Fig. 278. Te Kuiti Group limestone at the Awakino road tunnel.

erted pressure on the limestones.

Below the limestone is Aotea Sandstone (see page 139), which is creamy in appearance and contains layers of concretions and some poor shellfish fossils. Below that, under the grass, is Whaingaroa Siltstone, slumped as usual; and below that, west across the river on the Taumatamaire Hill, is Murihiku basement. It can be seen a few kilometres down the road in quarry

and roadcuts where it is made up of Jurassic sandstones, which are massive and jointed and contain a few barely discernible shellfish fossils.

The road to the coast continues through Mesozoic layers. About 7 kilometres past Awakau Road there is a quarry with poor casts of Triassic brachiopods in the sandstone, showing that the road has by then cut across the anticline layers and has reached the Triassic core.

142

THE BEGINNINGS OF THE GREAT DISTURBANCE

Auckland and North Auckland

In Northland also we can find the two familiar rock sequences: basement Torlesse and marine Cretaceous rocks with volcanics followed by white limestones similar to the East Cape-Marlborough rocks; and the sequence of uplifted land and coal measures and sea sediments we have just looked at in the King Country.

But their relative positions have been switched. The basement-coal measures series is in the east, from Whangaroa Harbour to the Hunuas, while the marine Cretaceous series is in the north and west (Fig. 279).

The question of how the sea-floor sediments arrived on the other side of the land mass has been hotly debated. Could the seaway have looped around the land, or could Northland itself have been twisted around on the shifting plates? The currently accepted explanation is almost as strange.

Imagine the two areas. Away to the east there was a sea floor, the northern continuation of the Cretaceous-Oligocene sea that formed the rocks of East Cape-Marlborough (Fig. 280). (Remember that a great fault runs through the North Island and that the present Hawke's Bay area

Fig. 280. Original pattern of land and sea.

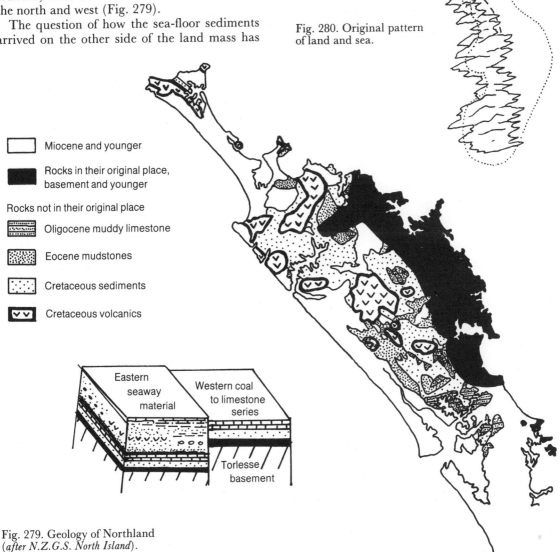

Miocene and younger

Rocks in their original place, basement and younger

Rocks not in their original place

Oligocene muddy limestone

Eocene mudstones

Cretaceous sediments

Cretaceous volcanics

Eastern seaway material

Western coal to limestone series

Torlesse basement

Fig. 279. Geology of Northland (*after N.Z.G.S. North Island*).

143

Fig. 281. Limestone at Waro.

may have once lain close to the Coromandel Peninsula.) To the west of that sea the whole Northland area followed the same pattern of events as occurred further south. There were coal measures, which have been mined at Kamo and Kawakawa, and the covering sandstones and mudstones were later draped with lime-

stone, visible at Waipu and Waro (Fig. 281) on the main road north of Whangarei. By the end of the Oligocene most of the Northland area was submerged under a warm sea, covered with layers of soft, limy ooze.

Then the first stirrings and upheavals jolted the ocean floor as the present collision course of the Indo-Australian and Pacific Plates began to be established. The sea bed buckled and lifted and great slabs of ocean-floor sediments rose up and slid on the soft ooze right across the area that is now eastern Northland. They came to rest right way up, in blocks measuring hundreds of cubic kilometres, up to 2.5 kilometres thick, their edges sheared and broken and smeared with clay. The various layers were not always in the same order; it seems they had separated as they moved and perhaps the younger, softer slabs had slid further (Fig. 282).

If this idea sounds far-fetched, the evidence has been found down drill-holes and in gravity surveys. Drill-holes show that the Cretaceous-Oligocene rocks in the west are actually resting on top of the same coal measures and younger sediments as those present in the east — older rocks over younger rocks. Gravity surveys show that the slabs of the Cretaceous volcanics are thin with no source areas beneath, and there is nothing to suggest that the volcanoes erupted on their present sites among the sediments that now surround them.

The volcanic slabs form large mountain ranges north of Kaipara Harbour, the Tangihua and Maungaru Ranges and others to the north-west. They are the most conspicuous of the transported ("exotic") layers. The higher parts of these Tangihua Volcanics are volcanic breccias containing glassy fragments and some

Layers of rocks slide downslope and perhaps separate, coming to rest in a different order

Fig. 282. The rearrangement of large blocks of rock in Northland.

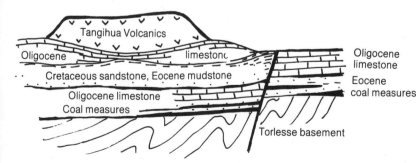

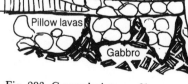

Fig. 283. General picture of layers in the Tangihua Volcanics.

144

sediments. Below is a mix of pillow lavas and massive basalt. The deepest part contains some pillows intruded by coarse, slowly-cooled gabbro and dolerite (Fig. 283). This mixture of rocks is fairly typical of a large volcanic centre on the ocean floor.

In places these volcanics contain copper sulphides, which were mined briefly at Parakao and Pakotai and prospected at Pupuke, near Kaeo.

The Tangihua slabs generally form steep-sided massifs above the surrounding softer Cretaceous and Eocene sediments which make up most of the rolling Northland landscape. The sandstone and siltstone contain some concretions and a few fossils — late Cretaceous *Inos* and some rare ammonites. These have been collected at the mouth of the Awapoko River west of Coopers Beach in Doubtless Bay and also at Batley and Bull Point on Kaipara Harbour, although nowhere are they easy to find.

The Eocene and Oligocene rocks in the west tend to be dark shales overlain by fine-grained, white, muddy limestone. Although widespread, they are part of the rolling, weathered landscape and are mostly seen in quarries and roadcuts, like the white banks near Opononi and Pakanae Beach on Hokianga Harbour. The limestone is quarried near Matakohe; from the quarry you can also see the limestones and Cretaceous sandstones south along Kaipara Harbour.

In much of western Northland the displaced blocks of sediment are so large that their displacement is not obvious. But there are also areas where smaller blocks of sediment of all ages are mixed up. Some of the first areas to be recognised as part of this "chaos" are at Onerahi Peninsula, east of Whangarei, where the foreshore is made up of blocks of many different sediments, all jumbled up, under the basalt flow which caps and protects the peninsula. A similar chaos-breccia can be seen on the south-east side of Parua Bay, a few kilometres to the east. Here it lies between sediments of late Oligocene and early Miocene age, which was the time of the original movement and the beginning of the basin in which the next group of sediments were deposited. Geologists are still unsure whether the Onerahi Chaos Breccia is a marginal part of the first great mix-up or a slump within the Waitemata Group, our next group of rocks.

Waitemata sediments and Waitakere volcanics

The rearrangement of Northland material in Waitakian time (the last age of the Oligocene) led to a new pattern of land and sea (Fig. 284). The old Torlesse basement and its coverings were uplifted to the north and east, and a basin slumped in the west and south as far as a ridge between Hamilton and Pirongia, which sepa-

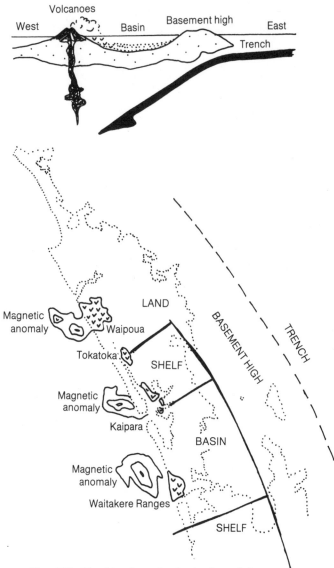

Fig. 284. Northland at the beginning of the Miocene (*after Hayward*).

145

Fig. 285. Paler conglomerate over dark Torlesse at Mathesons Bay.

Fig. 286. Ray holes at Mathesons Bay.

rated the northern basin from the Mahoenui basin to the south, where the widespread Miocene mudstones were accumulating (see page 140).

As the plate collision intensified, the downgoing plate produced a row of volcanoes out to the west, where the deeply buried volcanic masses offshore can now be detected only by gravity and magnetic anomaly patterns.

The Waitemata Group is all the layers of sandstone, siltstone and mudstone that built up in Pareora time as the seas deepened in the new basin. On the edge of the greywacke land mass to the east the seas lapped over the old rock, burying the greywacke with beach pebbles, shells, bryozoans, corals (Fig. 343, p.173) and algae. One buried shore can be seen at Mathesons Bay (near Leigh) below the memorial park (Fig. 285). In the conglomerate there you can find pebbles of greywacke surrounded by layers of algae which covered the pebbles as they rolled loose in the water as well as other fossils. As the seas rapidly deepened the sediments changed, and the upper cliffs are made up of layered sandstones and mudstones. Some basin-like hollows in the sands that are filled with coarser grit are thought to be feeding holes made by rays (Fig. 286).

The deeper-basin layers are visible from Mercer to Warkworth and Leigh in coastal cliffs and main road cuttings, especially in the headlands of North Shore and Auckland's East Coast Bays. Forams are the most common fossils, although in some areas fossil shells and some of the new tropical corals that reached New Zealand from the north have been found.

The sediments of the Waitemata Group contain many layers of volcanic ash from the row of volcanoes that were erupting to the west. With each eruption the whole area shook and trembled, and great storms brewed in the columns of ash from the eruptions. From storm flood or earth tremor the masses of unstable sediments on the shelf slid down to the deeper basin, making layers of very coarse rocks known as Parnell Grit, visible in Auckland at Parnell Point behind the baths. Broken shells were in the debris of the flow.

A really magnificent example of this kind of sediment flow is visible along the southern headland of Waiwera Beach at low tide. Just past the camp the layers of sandstone and mudstone in the cliff and on the shore platform show slumping and folding that took place when the sediments were still soft. At the turn of the beach cliff there is a 3-metre layer of extremely coarse material. The base is a conglomerate of volcanic boulders with zeolites in them (the boulders were the heaviest material in the flow and sank to the bottom first). The rest of the flow is made up of volcanic fragments, crystals and coarse sands, with a number of rolled-up lengths of soft mud and silt layers that had been caught up in the flow. Further along the beach the layers are cut by faults, and there is a classic anticline at the base of the cliff (Fig. 287).

Some of the chaotic flows within the Waitemata-Group sediments contain mixtures of many older rocks, as though some of the material transported from the east in the earlier disturbance had been knocked from a precarious resting place and carried away for a second time. As mentioned earlier (page 145) geologists are still sorting out the various areas of chaos in Northland, but one redeposited flow of this type is visible in cuttings on State Highway 1 on top of the hill about 11 kilometres south of Warkworth, where there is a layer of jumbled blocks within the regular layers of Waitemata sediments.

The Waitakere volcanoes erupted from three main centres: at Waipoua south of the Hokianga; between Dargaville and Hukatere on the Kaipara; and in the Waitakeres north of the Manakau (see Fig. 284). They may have begun erupting in the latest part of the Oligocene Epoch; but by Altonian time, 22 million years ago, the centres of eruption were moving east as piles of debris built up. At first the volcanic cones were constructed entirely under the sea, with the boulders and ash carried in various directions by wind and tide, reaching out and interfingering with the Waitemata layers. Finally, lava flowed from vents, making pillow lavas on submarine surfaces or building up small cones on the land (Fig. 288).

The two southern centres were most active at first. In the Kaipara the growing bulk of the submarine volcano, together with some uplift, meant that the eruptions soon flowed on to the land, and lignites and fossil wood were preserved in the ash shower material. For many years rockhounds have searched west of Tinopai for opalised wood — branches preserved by silica from the enclosing layers of ash.

Further north, near Tokatoka, many small masses of igneous rocks stand out above the limestone country rock. They are vents of scattered volcanoes, feeding channels up which the lava moved to a surface long since eroded away. Their shapes vary: lava moving up a round pipe forms a plug; but lava forced up through a narrow fissure forms a vertical wall cutting through surrounding layers — this is known as a dike, of which Maungaraho (Fig. 289) is a good example.

Fig. 287. Waiwera Beach: anticline at left by the figures, the fault slanting up at centre, and at right coarse conglomerate with rolled-up mudstone.

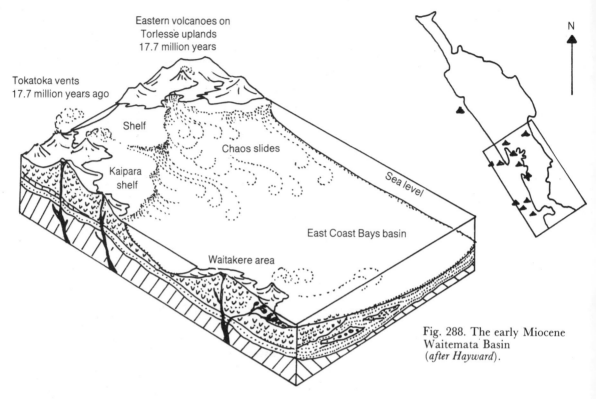

Eastern volcanoes on
Torlesse uplands
17.7 million years

Tokatoka vents
17.7 million years ago

Shelf

Kaipara
shelf

Chaos slides

Sea level

East Coast Bays basin

Waitakere area

N

Fig. 288. The early Miocene
Waitemata Basin
(*after Hayward*).

Fig. 290. Natrolite
from Arapohue
and analcite
from Maungaraho (bottom).

Fig. 289. Maungaraho.

The lavas are mostly basalts and andesites, with some dacite. In places explosions during the eruption broke the lava into pieces which formed a breccia, later cemented with calcite or zeolites.

Many of these volcanic outcrops, being much harder than the surrounding limestone, were quarried for road-metal or aggregate. In the small quarries, which are now overgrown, collectors found good zeolites (Fig. 290) in the cavities and cracks and breccias, and other minerals developed in the contact skarns with the surrounding limestone, although these are not easy to see.

At Todds Quarry, south of Arapohue Bush Camp Road, the rock is nephelinite, which looks like basalt but has less silica in its chemistry; it contains nodules of olivine. The quarry at Arapohue contains natrolite and analcite, and analcite also came from a quarry on Maunga-raho. Most of these volcanics have a potassium/argon radiometric age of 18 to 15 million years, dating them from Pareora to Southland time.

Further south the Waitakeres had been building up and extending to the east. On the north-west slope of the volcano lavas oozed up on to the sea floor, pushing forward tongues of congealing lava to make pillow lavas. The remains of these flows can be seen at Muriwai Beach, up the Waitea Road and off to the right, down into Maori Bay Quarry. (There is an excellent guide to this area published by the New Zealand Geological Society.) The wall at the back of Maori Bay contains layers of ocean floor muds and sands with ash fragments in them, and a thick conglomerate layer containing a few shells and corals which must have slid down into the deeper ocean from a boulder bank where the lava chunks had been rounded by the waves.

On top of these layers the cliff is formed from some of the finest pillow lavas in the country (Fig. 291). Among the smaller pillows are enormous lobes from feeder tubes which have cooled into radiating columns (Fig. 292); even the pillows have this radiating cooling pattern, giving the cliff the appearance of being formed from clusters of black flowers.

Under and over the pillows are layers of lava which flowed and cooled into a single row of columns. Such columnar jointing occurs in many volcanic rocks where the rock contracts around centres of cooling (Fig. 292). The lava

Fig. 291. Maori Bay, Muriwai: columnar basalt blocks on the beach; sediments in the cliffs below layers of pillow lavas.

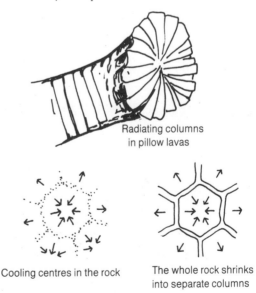

Radiating columns in pillow lavas

Cooling centres in the rock   The whole rock shrinks into separate columns

Fig. 292. Formation of columnar jointing.

here is an andesite very similar to basalt.

Above the quarry the pillows have skins of calcite where the lavas reacted with sea water.

On the north end of the beach are shallow-water sediments, sands and grits with layers of pumice which originally lay on top of the pillows. Soon after the eruption of the lavas the whole area began lifting, shallowing the seas and causing the waves to make patterns in the layers of sand, ash and pumice that were filling the basin. Later a fault (which runs through the beach car park) displaced the two areas, and the younger layers on the headland are now lower than the older pillow lavas.

As the volcano continued to grow, two lines of vents formed on its eastern slopes and they erupted ash showers and lava flows over the land to cap what are now the central Waitakere Ranges. Some of the rock in the outbursts was filled with gas bubbles, making coarse, porous scoria.

By this time the whole basin had been lifted up and the Waitemata sediments were complete. In the north-west, at Hokianga Harbour, rivers brought down great deposits of boulders to make a conglomerate some 300 metres thick which juts up as the South Head of the Hokianga.

Then the third volcano at Waipoua became active, covering the conglomerate with thick flows of basalt and forming the Tutamoe Plateau. Lavas from this volcano can be seen down Waimamaku Beach Road, where they cover a great, dipping layer of the conglomerate which is visible across the river. The road cuts through the lavas close to the beach. Most of the plateau is covered with kauri forest.

The Waipoua Basalts, on the west of the Waitemata basin, were the last to erupt; but by this time, some 15 million years ago, a row of volcanoes was forming to the east, from Whangaroa to the Coromandel.

Eastern volcanics

At the harbours of Whangaroa and Whangarei a series of andesite eruptions built up masses of breccia with some lava flows and dikes. Near Whangarei the breccias form Manaia Range (Fig. 293) and Bream Head Range; they can be seen close-up on the beach at Little Munro Bay, where the muddy limestone at the back of the beach is separated from the breccia on the rocks to the west by a grey dike of andesite with flecks of black pyroxene and white feldspar. The breccia is coarse, containing fragments of lava up to more than a metre wide.

At Taurikura Bay, opposite Marsden Point, a dike of andesite runs out into the harbour, forming a natural jetty (Fig. 294).

Fig. 294. Andesite dike, Taurikura Bay.

150

Fig. 293. Volcanic breccia on Manaia Range, Whangarei Heads.

▥	Kiwitahi andesite
▨	Whitianga Group Rhyolite, ignimbrite
▨	Waitawheta dacite
⌄	Coromandel Group Andesite, rhyolite
■	Paritu diorite

The Coromandel Peninsula

Although there are some areas of basement Jurassic greywacke, the bulk of the Coromandel is made up of Miocene and younger volcanics. See Fig. 295. At Colville the volcanics lie on river sands and coal measures, showing that the area was land when the volcanoes were active.

The volcanics have been divided roughly into two groups — the earlier, widespread Coromandel andesites, and the later Whitianga rhyolites — but in places we know both groups were erupting at the same time as each may occur under the other.

Most of the andesites have been greatly altered (see page 152), but fresh andesite is quarried on McBeths Road, between Hikuai and Whangamata, and it can be seen on the south side of Kuaotunu Hill, north of Whitianga, where the blue-grey roadcuts are patterned with small white zeolites, especially stilbite and heulandite, and associated calcite.

In the north of the Coromandel Peninsula rises the bulk of Mount Moehau, which is made up of diorite, the plutonic equivalent of andesite. It began to cool some 17 million years ago and has been quarried at Paritu for use as a building stone, also making the jetty from which it was shipped. Although called "Coromandel Granite" it has more iron-magnesium and more calcium in its chemistry than a true granite (Fig. 296).

The Whitianga rhyolites are mainly in the east. Some are at least 6 million years old and the younger ones some 3 million years. They are often flow-banded (banded by layers forming in

Fig. 295. Volcanics of the Coromandel area (*after N.Z.G.S. North Island*).

Fig. 296. Diorite, Paritu.

Fig. 297 Spherulitic rhyolite, Paku, Tairua.

Fig. 298. Mordenite, Puketui Hills.

the flowing lava). In some places the cooling lava crystallised with spheres of radiating clusters of feldspars and quartz, forming spherulitic rhyolite (Fig. 297). At the beach on the north side of Paku, at Tairua, there are many examples of spherulitic rhyolite on the rocks and beach pebbles. Other examples can be found in the Tairua River.

Whitianga rhyolite contains cavities with feathery tufts of the zeolite mordenite (Fig. 298), which is found in a roadcut up the Puketui Hills road, west of Hikuai. Once on a farm near Hikuai there was a hillside of grey, glassy rhyolite which contained tiny areas of precious opal. Common opal is more usually found, especially on the road to Hahei and on Great Barrier Island.

The alteration of the Coromandel andesites was caused by hot waters rising up through them — these waters carried metallic minerals in solution which were deposited in cracks and cavities.

There are two possible sources of these hydrothermal (hot-water) deposits: the magma itself or heated groundwater; possibly both were involved (Fig. 299). Some of the mineralisation

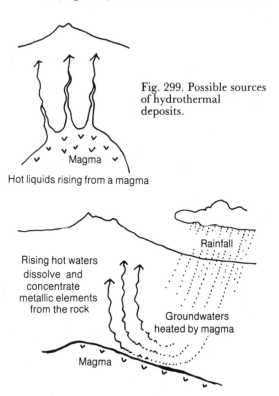

Fig. 299. Possible sources of hydrothermal deposits.

Hot liquids rising from a magma

Rainfall

Rising hot waters dissolve and concentrate metallic elements from the rock

Groundwaters heated by magma

Magma

at Mount Te Aroha has been dated at the 6-million-year mark, about the time of the formation of the eastern rhyolites.

As the hot liquids rose through the rock they deposited layers of crystals on the sides of the cracks through which they flowed. Sometimes the minerals built up to fill the whole crack and sometimes the cavities opened out and the crystals grew very large and perfect (Fig. 300). By studying the order in which the minerals grew on top of each other you can tell the changes that must have occurred in the hot liquids from time to time. The vein systems and pattern of cracks are often related to directions of faulting and movement in the area as a whole.

The deposits in the quartz veins of the Hauraki mining district provided our richest goldfield. Unlike the goldfields of the West Coast and Central Otago, nearly all the Coromandel gold occurred in veins, so the goldmines were mostly underground, following the quartz seams, with batteries to crush the ore and cyanide-processing plants to extract the gold. More than 35 million ounces of gold and silver came from the mines at Waihi.

The Coromandel has been a collector's realm for many years. Gold and silver ores and the associated crystals may still be found on old mine dumps, but they are now rare. Miners are again active at Waihi, Maratoto and Karanga-hake. Lead, copper and zinc were mined at Mount Te Aroha and occurred scattered in

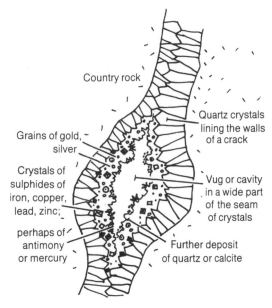

Fig. 300. Mineral deposits in a vein system.

other areas, such as up Tararu Creek in the Sylvia Mine.

Galena, the lead sulphide, is a heavy, grey, metallic mineral with distinctive cubic cleavage. Sphalerite, zinc sulphide, may also look metallic grey, but is more often brownish with yellow lights inside, like resin (miners once called it "gum ore"). Its cleavage runs in six directions, so it can be distinguished from blocky galena

Fig. 301. Tokatea Hill. A track leads down to the left below the road to a site for amethyst crystals.

pieces. The copper ore present is usually brassy chalcopyrite (Plate 8F).

Manganese ores, including pink manganocalcite, the carbonate rhodochrosite and inesite, a fibrous manganese silicate, were found at Waihi and Tararu Creek. Iron pyrite abounds, and stibnite is found up Waiotahi Creek near Thames, together with barite crystals.

Samples of the best crystals collected from the mines can be seen in the Thames School of Mines Museum, the Coromandel School of Mines Museum, local collections and the Auckland Museum. The mine dumps have been well picked over and at Waihi are being processed again.

The quartz family still provides good hunting. Pale amethyst may still be found at Tokatea Hill (Fig. 301), above Coromandel on the Kennedy Bay road, but there are many dangerous mine shafts scattered over the hill, so fossickers should be very careful. Old mines should be entered only by properly equipped expeditions. Many of the mining dumps contained quartz crystals. They are collected up Te Puru and Waiomu Streams and could be found in any Coromandel river.

Quartz in its cryptocrystalline form (crystals too small to be seen) is the agate and chalcedony that once was common along any Coromandel beach or river edge. The golden orange and red carnelian was especially prized; it was found weathering out of the hillsides and swamps along the western hills from Te Mata to Coromandel and beyond. Today in some forestry areas the streams still contain carnelian, but digging is strictly prohibited (Plate 8C).

Massive silica mixed with various impurities is known as jasper, which is also found in many varieties in the Coromandel. One impurity is plant material; a silicified swamp debris is known as "Manaia stone" from one locality. It often has a golden rind, white border and black centre and may contain stems, seeds and leaf fragments. It has been found in many rivers, near Coroglen and also on the west end of Otama Beach, past Kuaotunu. Petrified wood (Fig. 302) has been picked up in many rivers and beaches — Wilsons Bay, Amodeo Bay, Mill Creek near Whitianga, Coroglen and Rangihau Rivers and the Kauaeranga River behind Thames. The agatised wood from Table Mountain is magnificent, but now lies on forestry land, where digging is forbidden. Some of the

Fig. 302. Petrified wood in the Rangihau River.

wood from the Coroglen area is opalised with a colourful wood grain.

Once the Coromandel beaches and rivers yielded bags of quartz gemstones, but now such pieces are hard to find.

On the western side of the Hauraki depression there was a series of volcanic eruptions also in the late Miocene. They were from the Kiwitahi volcanoes running from Waiheke Island (the oldest, being 11 million years), south to Maungakawa, which is about 5.5 million years old. These were all fairly small flows of basalts and andesites with crystals of feldspar and pyroxene in a fine-grained groundmass.

South Island volcanoes

While the Coromandel volcanoes were active, other volcanoes broke out in the South Island; and whereas the Coromandel lavas were andesites and rhyolites, those in the south were mainly varieties of basalt.

In the Miocene small eruptions produced the low hills north of the Waimakariri near Oxford, Burnt Hill and Starvation Hill. To the south of

Coalgate the Harpers Hills are capped with a layer of basalt and tuffs.

These were the little ones. The great volcanoes formed Dunedin and Banks Peninsula and their beautiful harbours.

Dunedin

After its uplift the basement schist here had been covered by the familiar coal seams overlain by sands, muds and greensands. They show up very well at Fairfield where at the Walton Park Sand Quarry (Plate 5F) the top marine layers are being stripped off for the mining of the silica sands and pebbles above the coal measures. The coal in the pit to the south-west is Cretaceous, and Cretaceous-age rocks reach up into the middle of the greensands, where the concretions sometimes contain ammonites.

The sloping marine layers on top contain thin seams of clays which, when wet, act as a lubricant on which the overlying rock may slide, with the disastrous results that occurred at Abbotsford.

Shallow seas still covered the Dunedin area at the end of Southland time, when the first eruptions began near Portobello, building up a cone of ash and lava flows of trachyte, a fine-grained rock like andesite but containing more soda-rich feldspars.

Three main periods of volcanic activity followed the initial eruption. (See Fig. 303.) The first was again centred on Portobello, with the

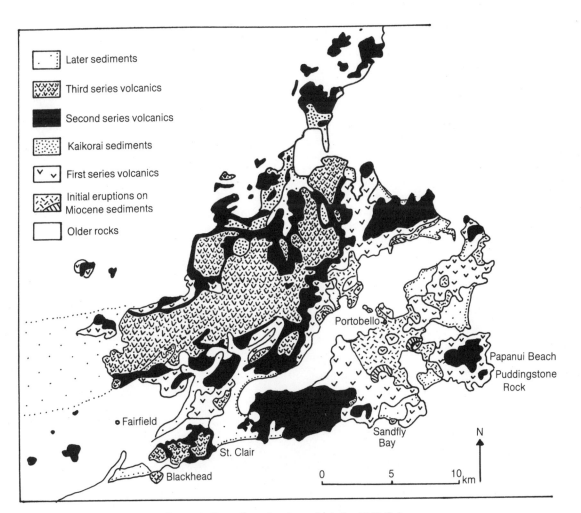

Fig. 303. Dunedin volcanic rocks (*after McKellar*).

155

Galaxias sp.

Fig. 304. Fish fossil, Kaikorai Valley.

Fig. 306. Calcite, Otago Peninsula.

flows spreading some distance on to both land and sea. When they ceased, the new land began to erode; trees grew and leaves fell into the mud by pools where tiny freshwater fish swam and were buried, leaving remains (Fig. 304) which are now found in Kaikorai Valley in south-west Dunedin and in Leith Stream.

Lavas from the second volcanic period spread much further, building up high land on Cargill and out to the present limits of the peninsula. These rocks, too, were covered at places with a layer of sediments. Eruptions of the third period occurred mainly in the north and west, especially along the Cargill-Mihiwaka ridge with flows that spread across Flagstaff to reach the Taieri Plain. Other eruptions formed Saddle Hill and Scroggs Hill to the south and many smaller hills north to Palmerston.

The eruptions at Dunedin began about 12.5 million years ago, just after the Waipiata eruptions, and lasted some 4 million years. By this time faults were shifting the blocks of land up and down, forming the Taieri Plains and defining the harbour basin. The shield volcano

was distorted to the west and also deeply eroded.

The lavas in Dunedin vary from basalt to andesite, trachyte and phonolite — a kind of trachyte containing nepheline. In the naming of rocks there is no end, but to distinguish these rocks you need a microscope. At Port Chalmers there is a volcanic breccia produced by explosive eruptions which tore up bits of the country rock (older rock through which the volcano erupted) and brought up rocks from deep underneath; it contains syenite and gabbro as well as schist.

As the lavas cooled they contracted into columns. The columnar jointing shows up very well at St Clair (Fig. 305), along the walkway from the end of Second Beach Road. The cliffs to the south were built up from at least three flows

Fig. 305. Cliffs south of St. Clair Beach.

of basalt, each of which cooled in rows of columns. In the banks at the beginning of the path are visible layers of volcanic fragments and ash.

Cavities in the basalt contain calcite (Fig. 306), aragonite and zeolites such as analcite and natrolite. Aragonite has been found at Sandfly Bay. At Papanui Beach near Cape Saunders the boulders along the beach contain calcite clusters and some aragonite. (See Plates 6A and D.) On the other side of the headland at Puddingstone Bay the cliffs contain seams of opalite, patterned and spotted with white, red, green and blue-white. The seams are bedded in very solid basalt and collectors must work very hard to extract anything larger than chips. They use ropes to help them down the cliffs onto rocks which are covered at high tide.

At Blackhead Quarry the basalts contain calcite and zeolites, especially natrolite.

Banks Peninsula

This great shield shape was built up by two overlapping volcanoes which erupted onto an island of greywacke.

The early eruptions which occurred in the Cretaceous Period were of rhyolite and andesite. The locations of these rocks on Banks Peninsula are shown in Figs. 307 and 308. In McQueens Valley the andesite is very similar to that in the Rakaia Gorge, containing agate and zeolites.

The first Miocene eruptions covered Torlesse rocks on the foreshore of Governors Bay with grey andesite, while rhyolite domes grew at Charteris Bay and Gebbies Pass, where they contain tiny dark crystals of quartz. Under the lowlying rhyolites are beds of sandstone which was used as a building stone, taken from Blatchford's Quarry in the Bradley Park Estate.

Some 12 million years ago the Lyttelton volcano began to pour out basaltic lavas with occasional outbursts of ash and small pebbles. On Summit Road, just over 3 kilometres west of Gebbies Pass, the road cuts through tuffs and

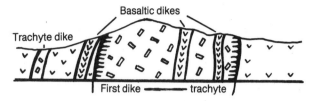

Fig. 309. Dikes on Summit Road. (*Geology Dept., Canterbury University.*)

blocky basalt on top of eroded and weathered rhyolite (Plate 6E), showing the base of the flows that eventually built up the cone to some 1,500 metres.

After that the cone lifted (it was perhaps pushed up by the rising mass of Akaroa magma) and it began to split open with fine cracks radiating out from a centre at Charteris Bay. Lava rose up through the fissures in a swarm of dikes, many of trachyte, which can often be seen along Summit Road. One well-known group is near the car park at the lookout, 800 metres south-west of the Sign of the Kiwi at Dyers Pass. Here several dikes can be distinguished, each with a chilled margin where the lava cooled against the cold older rock (Fig. 309). Across the harbour the ridges of the dike swarm above Charteris Bay show up well.

About 9.7 million years ago lava began flowing on the Lyttelton rim and built up Mts Herbert and Bradley as the volcanic eruptive centre moved south-east towards Akaroa Volcano, which began erupting from the south-east about 9.5 million years ago and continued to grow until it overtopped the Lyttelton cone.

The Akaroa volcano also is basaltic, but in the centre at the northern shores of the harbour and by Onawe Peninsula there are underlying flows of trachyte and many dikes. On the end of this peninsula crystalline gabbro and syenite are present, as well as ash layers and breccias (Fig. 310). The peninsula is now an Historic Reserve of the Lands and Survey Department; look for the Maori pa site on the heights.

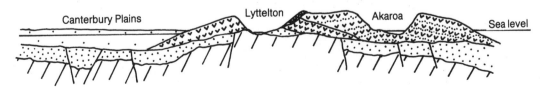

Fig. 308. Cross-section of Banks Peninsula. (*Geology Dept., Canterbury University.*)

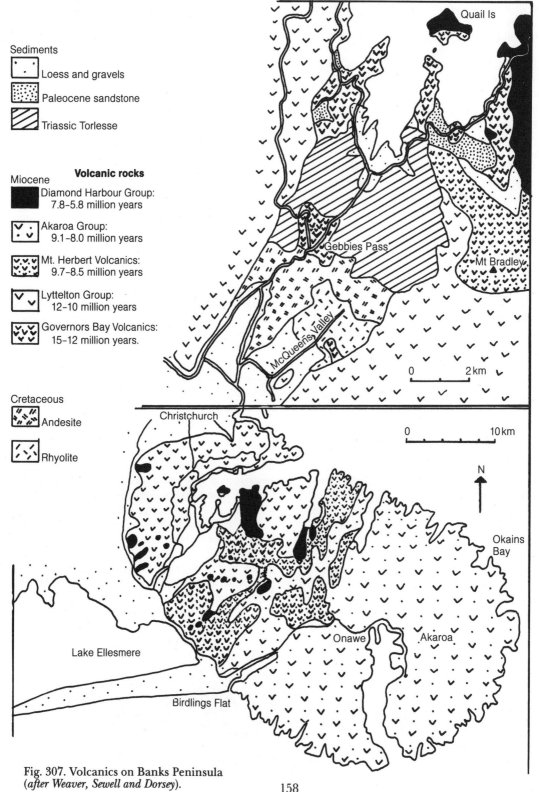

Sediments

Loess and gravels

Paleocene sandstone

Triassic Torlesse

Volcanic rocks

Miocene

Diamond Harbour Group:
7.8–5.8 million years

Akaroa Group:
9.1–8.0 million years

Mt. Herbert Volcanics:
9.7–8.5 million years

Lyttelton Group:
12–10 million years

Governors Bay Volcanics:
15–12 million years.

Cretaceous

Andesite

Rhyolite

Fig. 307. Volcanics on Banks Peninsula
(*after Weaver, Sewell and Dorsey*).

The last eruptions came from a vent near Mount Herbert between 8 and 5.8 million years ago, when a basalt containing olivines flowed down the slopes into Lyttelton crater and spread across to Quail Island. Smaller flows erupted near Ahuriri and Halswell, where the basalt has been quarried.

When the eruptions died down, streams began to cut channels in the slopes. On the Lyttelton volcano one stream on the north-east side cut deeper until it reached the crater. With the rise in sea level after the last polar ice melt, the sea entered the crater, which continues to be cut back by erosion. Akaroa Harbour was formed in the same way.

As at Dunedin, the lavas here contain cavities with zeolites (Fig. 311), calcite and aragonite. In Lyttelton Quarry and the boulders taken from it the lavas have cavities with tiny chabazite crystals and garronite balls, as well as calcite and some aragonite. Out at Okains Bay various forms of aragonite (Fig. 312) have been found in the cliffs.

The New Zealand Geological Society has published a complete guide to the Lyttelton and Akaroa volcanoes (see page 218).

SEDIMENTS IN THE LATE MIOCENE — THE KAIKOURA UPLIFT

All the volcanic activity in the Miocene was part of a renewal of crustal movement. After the relatively peaceful interlude of the Eocene and Oligocene, New Zealand was mobile again, and the pattern of sediments around the land shows it.

During the quiet time in the Early Tertiary very little sediment was washed off the low-lying land, and the Oligocene limestones were built up over the centuries from the remains of shellfish. But after that the entire sea shelf became covered by vast amounts of silts and sands which came down from a land that was raised above the sea again and rapidly eroding. The layers contain lenses of coarse pebbles with rock fragments and minerals that are still fresh, unaltered by long exposure to the weather. These could have come only from nearby land that was recently uplifted, probably in a shape that foretold the line of the present main ranges in both islands.

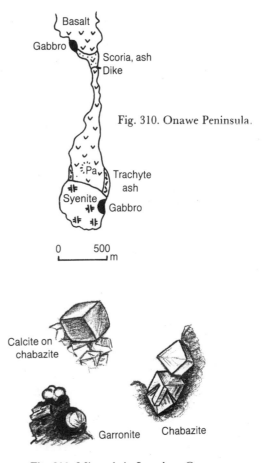

Fig. 310. Onawe Peninsula.

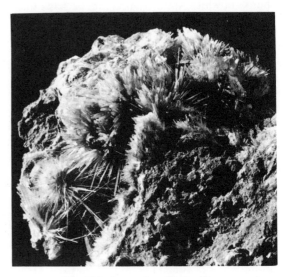

Fig. 311. Minerals in Lyttelton Quarry.

Fig. 312. Aragonite, Okains Bay.

159

Fig. 313. Great Marlborough
Conglomerate near
The Shades Station.

Fig. 314. Detail of cliffs at the south
end of Awakino Beach, showing the
top of the Mokau Sandstone and the
thin layer of conglomerate at the base
of the Mohakatino layers.

In Marlborough in the Early Miocene (Pareoran-Altonian times) the rivers carried loads of boulders down channels cut far out on the sea shelf, building up a submarine fan. Layers of this Great Marlborough Conglomerate can be seen up Deadman Stream; and on State Highway 1, if you look north near The Shades station, the cliffs contain thick bands of conglomerate tilted down to the west (Fig. 313). Just south of Kekerengu the cliffs in the roadside and railway cutting are studded with great chunks of white Amuri Limestone, perhaps torn up from the sea floor in the surge of bouldery debris.

In northern Taranaki the flexing of the land was felt in the early Miocene when the Mokau Sandstones were being laid down on top of the Mahoenui Mudstones. At the same time as the Waitemata uplift, the area that is now North Taranaki was pushed up out of the sea to become land. On the new coastal plain peat swamps developed among river gravels and silts, where rivers built out a delta against the encroaching sea. Today there are several open-cast coal mines near Waitewhena in the Maryville Coal Measures. Unlike the earlier Waikato coals, these both lie on and are covered by marine sands and muds, showing that there was only a brief period of uplift in the area during Pareora time. The Maryville Coal Measures are the filling in the Mokau sandwich,

with Mokau Sandstones and mudstones on top and bottom — the same sandstones as those at the Awakino Tunnel.

Nearer the coast they are topped by the younger (Southland-time) sandstones and mudstones of the Mohakatino Group. These can be distinguished by bands of volcanic material similar to those in the older Waitemata Group. Like the earlier volcanic material, the ash is assumed to have come from volcanoes erupting offshore to the west where there is another magnetic anomaly off the Awakino coast.

The change from Mokau muddy sandstones with concretions to the overlying Mohakatino rocks shows up at Awakino Beach (Fig. 314) at mid to low tide. At the base of the cliffs to the south is mudstone, a continuation of the mudstone in the river with its layers of concretions. Directly above it is a grey sandstone containing shell fragments, brachiopods, echinoid spines and small corals. Over this is a thin band of conglomerate, including concretions with worm borings, showing a break in sediment build-up and the start of the new group of rocks. The cliffs above are formed of Mohakatino layers of sands containing volcanic crystals and rock fragments.

Further down the Taranaki coast the layers are younger (see Fig. 269, p.138) and the soft, grey cliffs are typical of the marine layers forming all around and over New Zealand at this time.

The Taranaki Series — last of the Miocene

The type section for the Taranaki Series is the sediment beds on top of the Mohakatino Group, represented along the coast for some 16 kilometres south of the Mohakatino River to Tongaporutu township and then inland up

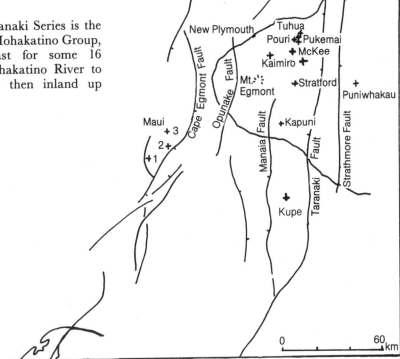

Pliocene mudstone

Pliocene sandstone

Urenui Siltstone

Mt. Messenger Sandstone

Mohakatino rocks

Mokau Sandstone

Mahoenui Mudstone

K Kapuni Group

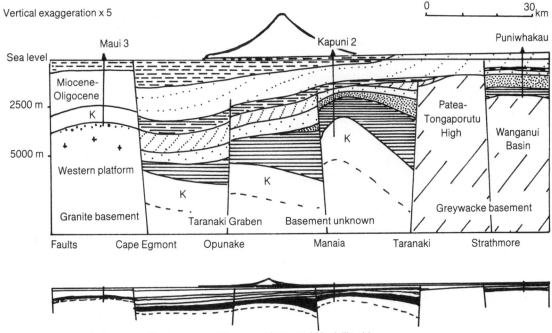

Vertical exaggeration x 5

If the vertical exaggeration is removed the proportions would look like this

Fig. 316. The rocks under Taranaki (*after McBeath*).

161

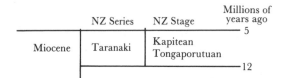

	NZ Series	NZ Stage	Millions of years ago
			— 5
Miocene	Taranaki	Kapitean Tongaporutuan	
			— 12

Mount Messenger (Fig. 315). The lower layers are predominantly sandstones, the upper layers are siltstones. They all contain some ash beds, some layers of concretions and some large fossils, together with abundant forams which define the Tongaporutuan Stage. (Kapitean-Stage fossils are much rarer; the Kapitean Stage is discussed later [page 168] when we return to the West Coast and Kapitea Creek, near Greymouth.)

All through central Taranaki and to the east similar beds occur, with some shell reefs and limestones, the youngest of the sediments filling the Taranaki Basin. This old basin, extending far west of the present land, is filled with the same layers as those tilted up to the surface as far north as the Waikato and contains our most likely reserves of petroleum fuel (See Fig. 316).

Fig. 315. Tongaporutuan sandstone on Mount Messenger.

Oil and gas in the Taranaki Basin

Some swamps contain gas, and peat and coal also produce gas, which is one of the great dangers of underground coal mining. Oil is another compound of carbon and hydrogen produced from the decay of plant material. (Overseas, most commercial oil deposits come from decayed marine organisms, although Australian Gippsland oil is a product of coal measures.) Coal-forming swamps on the Cretaceous and Eocene land between the Waikato area and Golden Bay have already been described. In the Taranaki Basin these have been deeply buried under later sands and silts and are the source of the oil and gas found in Taranaki, the Kapuni Group of rocks.

Unlike coal, oil and gas do not stay in the layers in which they form. Once the heat and pressure of the overlying rocks have cooked the plant material, each tiny, fluid molecule of gas and oil will tend to make its way up towards the surface of the ground where it will escape in a gas or oil seep.

To be useful for industry, oil and gas must be trapped underground and held there under pressure so it can flow or gush up a drilling well. The two substances must collect in a porous reservoir rock, such as a coarse sandstone, or a more dense rock with fractures, and the spaces in the rock must be connected so the hydrocarbons may move freely once the drilling pipe enters the reservoir (that is, the rock must be permeable as well as porous — see Fig. 317).

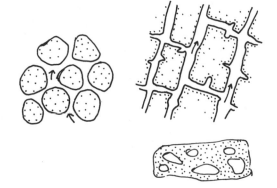

Fig. 317. Porosity in rocks. Top left, sandstone with pore spaces. Top right, hard rock with cracks. (Both of these are both porous and permeable.) Bottom, pumice, which is porous but not permeable, because the spaces are not connected.

The hydrocarbons have to be held down in the layers of the reservoir, usually by a layer of dense, impermeable rock, such as a mudstone. This layer must be shaped so the oil collects in a pool. The simplest shape is the uplifted dome of an anticline, but others occur where a fault closes off a reservoir or where lenses of porous material occur in finer materials such as coral reefs or sandy channel fillings in an area of mudstone (Fig. 318).

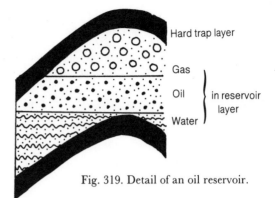

Fig. 319. Detail of an oil reservoir.

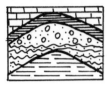

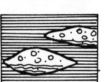

Fig. 318. Common shapes of oil reservoirs.

"Oil pools" (Fig. 319) do not occur in open cavities, but fill pore spaces in a rock. Generally, the gas separates out and fills the highest part of the reservoir while the heavier oil and any dissolved waxes remain beneath; the pore spaces below that are usually filled with water.

The Taranaki reservoir rocks are sandstones like the Aotea Sandstone, and the trap-rocks are dense, brown shale or mudstone. On top of the Kapuni Group of rocks is a fine-grained, deep-water equivalent of the Te Kuiti Limestone; it is also Oligocene in age (Fig. 320).

Geologists searching for oil first locate underground structures likely to trap oil, given that they believe suitable source rocks are in the underground layers. In a seismic survey they fire a series of explosive charges and record

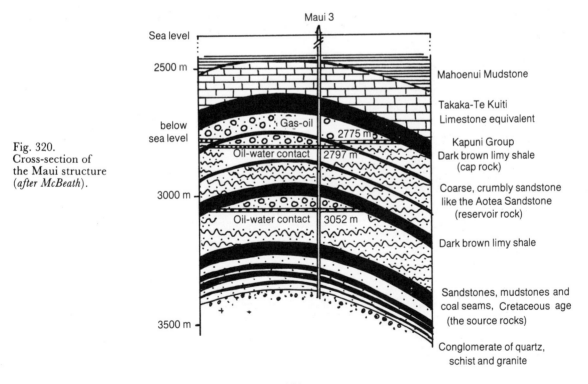

Fig. 320. Cross-section of the Maui structure (*after McBeath*).

163

reflected sound waves bouncing off the layers below (Fig. 321). Harder layers show up as darker patterns.

The only way to test whether oil is trapped under a dome is to drill, which is very expensive. There have been a number of wells drilled around New Zealand that are "dry" — that is, they do not contain oil or gas in commercial quantities. However, every well drilled provides us with a better picture of the layers below the surface.

At present three areas produce mainly gas and condensate — Maui, Kapuni and Kaimiro. (Condensate is oil dissolved in gas and occurs when gas is under pressure in the earth. It condenses into a liquid when it reaches the surface.) McKee oil field is in production, with Tariki and Ahuroa, and offshore Kupe is being tested.

The shape of the Taranaki Basin had been controlled to some extent by movement along the Cape Egmont and Taranaki Faults, which from time to time deepened the central portion and lifted the ridges on either side to form a graben like the contemporary Hauraki Valley graben. In the graben thicker layers of sediments built up. To the east the high block (known as a horst) must have been land for part of the time, judging by the layers that are missing. However, the Taranaki Basin never extended much past Wanganui in the south and Ohakune in the north. To the east a wide belt of land, an early Ruahine upland, rose to cut off the waters covering the present East Coast.

The East Coast Miocene rocks
From East Cape to Palliser Bay patches of grey sandstones and mudstones of the Miocene sea floor can be seen by travellers. In the Miocene the old Cretaceous-Eocene shelf continued to fill with the thick sediments that still blanket the area and are now cut by erosion in river valleys

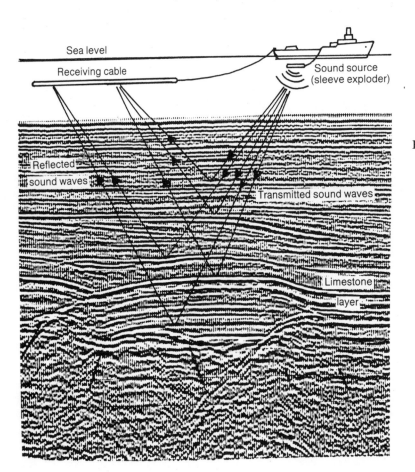

Fig. 321. Seismic survey at sea.

and sea cliffs. At many places Tongaporutuan Stage large fossils have been found in these sediments, but Kapitean Stage macro-fossils are rare, and the younger layers are mostly identified from forams.

At Te Araroa white fossil shells in the cliffs east of the town (Fig. 322) are fragile and break easily; those the rain has washed out may be harder and intact, but they are more difficult to find. The rows of broken *Cucullaea* in the cliff serve only to frustrate the collector.

Fig. 324. Makorori Beach, Gisborne.

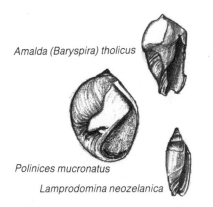

Amalda (Baryspira) tholicus

Polinices mucronatus

Lamprodomina neozelanica

Fig. 322. Southland Series fossils at Te Araroa.

Whole *Cucullaea* in pairs may be found with a search on the headland beach past the Tokomaru Bay wharf. Also present are whale bones and a species of small heart urchin. This echinoid (Fig. 323) lives in deep, soft muds and often is the only fossil larger than foraminifera found in these sediments.

Maretia sp

Fig. 323. Heart urchin from Tokomaru Bay.

At Makorori just north of Gisborne the beach is ribbed with alternating layers of sandstone and muddy siltstone of middle Miocene (Southland) age (Fig. 324). They show graded beds and ripple marks from currents carrying in the sediments, just like those in the older Torlesse beds, but these are softer rocks.

Across Poverty Bay the white cliffs of Young Nicks Head are made of massive Taranaki

(Kapitean) siltstones, but further south on the main road the roadcuts up the Wharerata Hill are again in banded Southland sandstones and mudstones. The graded beds show up well in the roadcut by Wharerata Walkway.

As you travel south to Morere the banding disappears and the younger Taranaki rocks are massive grey siltstones; at Moumoukai the tabletop bluff to the west is capped by even younger rocks, with Pliocene limestone on top.

South of Hawke Bay the old seaway continued to fill with Oligocene and Miocene sediments. The warm Oligocene seas swarmed with microscopic life whose shells increased the amount of lime in the mudstone, although true limestones are rare here. The white cliffs near Weber (Fig. 325) are typical of the Oligocene rocks of this whole area.

Fig. 325. White muddy limestone at Weber.

Fig. 326. Maunsell's Taipo, Tinui.

All these rocks are Southland in age, middle Miocene.

Fig. 328. Limestone in the disused quarry up Tinui Valley Road.

Then earth movements compressed this region also, buckling it so that long, narrow troughs formed in the former shelf, into which poured the same grey sands and silts as those found all around the coast. The Miocene layers varied from place to place, sometimes being a massive influx of sands, sometimes alternating layers of sand and silt or mud.

Massive soft sandstones topped by younger limestones form the cliffs of Cape Turnagain (Plate 7D). Hardened layers of sandstone hump their hogbacks into Maunsell's Taipo near Tinui (Fig. 326), east of Masterton; yet further down the road to the coast, at Whakataki, the rocks form alternating layers of sandstone and siltstone (Fig. 327). Inland in the Tinui Valley you can see thick sandstone and layers of a creamy limestone containing pinkish algae and black forams (Fig.328). It is a very decorative stone indeed.

Fig. 327.
Whakataki Beach.

About 10 million years ago, in early Taranaki time, there was further buckling of the crust. Some basins were just tilted and others were raised right above sea level so that a new land formed a barrier island in the eastern sea. At the south end the Aorangi (Haurangi) Mountains lifted their bulk.

At the southern end of these mountains a flood plain built up, with logs of wood among the river boulders. Fossil pollens show the flood plain was building up about seven million years ago. Today this area is eroding out and forming badlands, known as "The Pinnacles" (Fig. 329), part of the Haurangi Forest Park up the Putangirua Stream in Palliser Bay.

As the land warped again the seas moved in and the gravels were covered with layers of grey sea-floor sands and silts, with shells in them (Fig. 330). Whalebones found here have been surrounded by coprosma leaves, so the sea must

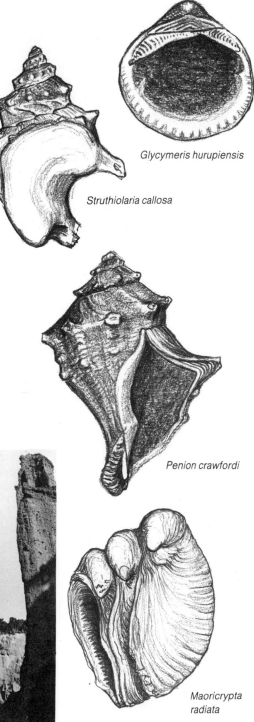

Glycymeris hurupiensis

Struthiolaria callosa

Penion crawfordi

Maoricrypta radiata

Fig. 330. Tongaporutuan Stage fossils in the Putangirua Stream.

Cominella hendersoni

Struthiolaria praenuntia

Fig. 329. The Pinnacles, Palliser Bay.

Fig. 331. Gypsum rosettes, Hurupi Stream.

have been fairly shallow. In the Hurupi Stream the sandstones also contain gypsum rosettes — radiating crystals of calcium sulphate, enclosing sand and small shells (Fig. 331).

Further east along the coast the tilted slab of rock called Kupe's Sail contains brachiopods and echinoids, sand dollars and many bryozoans in a limestone, showing that this area, too, was a shallow sea in Taranaki time.

In the South Island, at Weld Pass on State Highway 1, travellers can see the same sort of pinnacles eroding from gravels of the same age.

Down on the West Coast the basins persisted from earlier basins in the same area. South of Karamea in the Kongahu area there are thick marine sands and silts. Inland at Murchison the sea sound had become mostly river valley, but on the coast from Westport to Hokitika marine siltstones and fine sandstones of a generally blue-grey colour are found containing a few beds of fossils as we saw when looking at earlier West Coast rocks (page 115).

In North Westland these Taranaki mudstones were the "Blue Bottom" of the goldminers — the impervious beds underlying the goldbearing gravels (Plate 7A). Goldmining relics are common. Near Callaghans on the back road between Stafford and Dillmanstown the rusting ruins of a dredge sit in its pond. In this area of the Waimea Forest the Forest Service has cleared some of the old mining tracks and walkways as the Shamrock Creek Amenity Area and made a picnic ground at the Callaghans township site, now marked only by a house foundation and some fruit trees. Below the fence by the picnic area is a deep, narrow cut made as a drainage channel for the miners' sluicing area. Overgrown footpaths in the bush lead to many smaller cuts and views, some hazardous;

there is a safe view of the cut and the private workings of a gold claim some 15 minutes' walk up the Goldsborough Track which may be followed for a four-hour walk to Goldsborough, 3 kilometres from Stafford.

Along the cut geologists took samples for forams, since large fossils were rare here, and found a sequence that defined the upper Taranaki Stage, the Kapitean. The locality is ideal as a type section since good forams can be taken from the channel at measured intervals and the cut is so deep that only the dedicated would venture down.

In Kapitea Creek itself the blue mudstones contain a greensand bed and some large fossils (Fig. 332), typical of the Kapitean Stage, but in general fossils of this age are rarely found in New Zealand. The end of the Taranaki-age rocks is usually determined by the beginning of those of the next epoch in time, the Pliocene.

Sectipecten wollastoni

Fig. 332. Kapitean Stage pecten from Kapitea Creek.

The Wanganui Series — Pliocene-Pleistocene

Travelling south-east from Mount Egmont we move across the last great area to be covered by the sea, where changing patterns of sediments record most of the events of the last 5 million years. The Wanganui and Hawera Series bring the time scale up to the present, covering the end of the Tertiary and the whole of the Quaternary Periods. The boundary between the

Epoch		NZ Series	NZ Stage		millions of years ago
Quaternary	Pleistocene	Wanganui	Castlecliffian (Okehuan) Nukumaruan (Hautawan)	warmer cooler warm cold	
					1.63
Tertiary	Pliocene		Waitotaran (Mangapanian) Waipipian Opoitian		
					5

two is taken from the appearance of the major effects of the Ice Age.

The term Quaternary in New Zealand refers to the time when the present landscape was being shaped by ice and its melting, by volcanic eruptions and earth movement caused by shifting plates.

As in the Arnold and Landon Series, the type section for the oldest Wanganui stage is far away, in the hills behind Opoiti near Wairoa (Fig. 333). The boundary between the Miocene rocks and the Opoitian layers can be seen about 40 kilometres up Mangapoike Valley Road from Frasertown. The views make the trip worthwhile. Past the Makaretu Stream bridge the river falls in a series of broad steps over Opoitian mudstone.

On the summit above the junction of Makaretu Stream and Mangapoike River you can look west through the gorge towards Opoiti and the Wairoa Valley. The limestone here marks the top of the Opoitian layers and slants down to the west. It is covered by a second, thin limestone containing Waipipian fossils, and over that is some 30 metres of siltstone capped by a hard layer of volcanic ash (tuff) some 5 metres thick.

Further along the road the lowest Opoitian sandy limestone angles up across the road with a sharp overhang over the mudstone underneath. The limestone here is mostly composed of barnacle fragments with some large pectens or fanshells. In the past the limestone was taken as the base of the Opoitian Stage, but forams in the mudstone are also Opoitian in age.

The Pliocene-Miocene boundary can be seen

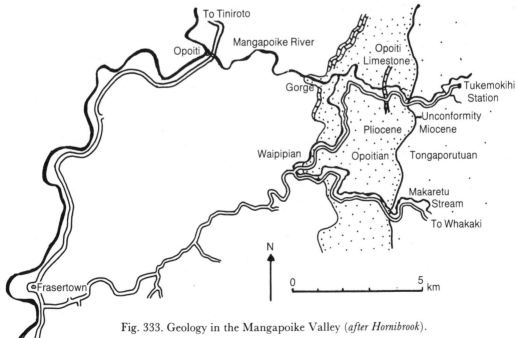

Fig. 333. Geology in the Mangapoike Valley (*after Hornibrook*).

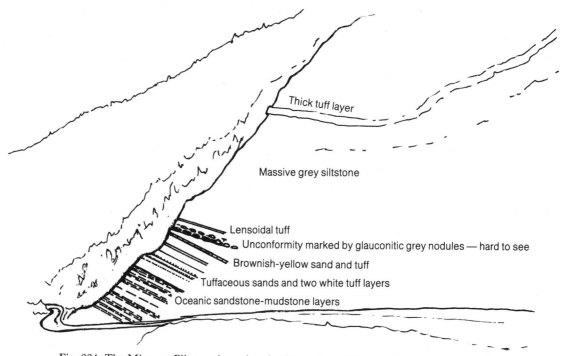

Fig. 334. The Miocene-Pliocene boundary in the roadcut, Mangapoike Valley. See Plate 7G.

in the cliffs where the road descends to the river and levels out. About 3 metres above the road the angle of the rock layers changes, showing that there was a period of time between the laying down of the two deposits during which the land was tilted. Since the angles do not match, the contact zone is called an "angular unconformity" (Fig. 334). We have already looked at these changes of angle, especially where younger rocks lie on the Torlesse basement.

Here the unconformity is only 3 degrees and the time gap between the layers is less than 1.5 million years, judging by the missing foram sequence. The Miocene (upper Tongaporutuan Stage) rocks form layers of mudstone and sandstone; the Pliocene rock above is a greenish mudstone with a layer of hardened rock at its base. Large fossils are rare; all the dating was done with forams.

Below the unconformity are two thin, white bands of volcanic ash (tuff); about 15 metres above it is a hard, yellow ash layer about a metre thick. Ash beds are conspicuous all

through this area. Ash fell in thick layers down through the sea on to the outer shelf, which became overweight and slumped down into the depths. Some of the ash layers have graded beds and ripples such as those in the Torlesse greywackes, and there are chunks of siltstone that have been ripped up and incorporated in the ash.

These tuff layers are common in the Taranaki-age rocks here and occur less commonly in the Wairarapa. They are all of rhyolite ash. Since the only such volcanoes erupting at that time were in the Coromandel Peninsula and further north, the large number of thick tuff beds suggests that the two areas were once closer together.

Over the ranges in the Wanganui basin the shallow Opoitian sea stretched between the Taranaki ridge and the rising Kaimanawa-Ruahine Ranges. On a clear day travellers looking north by Oroua Downs or Marton can see the former extent of the seas in the level plains below the later volcanoes (Fig. 335). Seen from east of Hunterville, Ruapehu seems to

Fig. 335. The Pliocene sea from the south (the volcanoes came later).

170

Fig. 336. East of Moawhango a farm track runs between the Torlesse rocks below and the Pliocene limestone covering them.

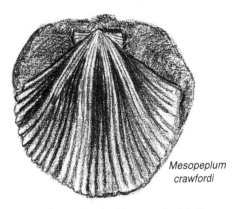

Mesopeplum crawfordi

Fig. 339. Pecten in the Mangaweka Mudstone.

stand on a pedestal of valleys cut in soft, sea-floor muds.

The northern edges of the sea are marked in the beach sands and gravels which fill old valley mouths near Moawhango. Where the Taihape-Hastings road dips down to the upper Rangitikei Valley the old Torlesse rocks are covered by sands and shell beds left by the encroaching sea (Fig. 336).

Layers of sandstone and limestone form the flat-topped hills south and east of Waiouru, reminding travellers on the Desert Road that this area was under the sea less than 5 million years ago (Fig. 337). Near Hihitahi limestone blocks tumble down beside the highway. There are several layers of harder limestones on the hillsides below the layer that forms the flat, upper surface. The lower layers are marked by lines of shrubs growing on the vertical faces of the well-watered and fertile limestone.

Further south at Taihape the water deepened, perhaps to 100 metres, and the sea was floored with blue-grey muds, forming the papa mud which is such a headache to road builders today.

A brief shallowing of the sea produced the brownish sandstone near Utiku with its rows of knobbly concretions; they are very hard to crack to find the small shells inside.

The seas deepened again and covered the sandstone with more muds which filled the basin south past Mangaweka, where the Rangitikei and Turakina Rivers have cut their smooth, grey walls (Fig. 338). Boulders containing large fanshells (Fig. 339) have been found in the river at Mangaweka.

All the material that has filled the Wanganui basin has been cut along the coast from Taranaki to the Manawatu River, exposing the layers (Fig. 340). The stages of the Wanganui Series younger than Opoitian can be followed by going east past the townships and rivers whose names they bear, although in some places river flats and sand dunes hide the rocks on the coast and better examples can be found inland.

Fig. 337. Pliocene limestone bluffs east of Waiouru from the Desert Road.

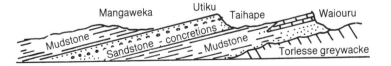

Mangaweka Utiku Taihape Waiouru

Mudstone Sandstone - concretions Mudstone Torlesse greywacke

Fig. 338. Pliocene layers in the Rangitikei Valley.

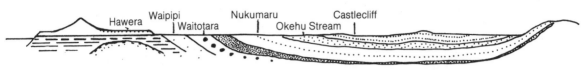

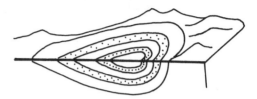

Fig. 340. Pliocene to Pleistocene layers in the Wanganui Basin.

To the west in Taranaki Opoitian siltstones form the sea cliffs at Ohawe Beach (eastern end), Waihi Beach and Hawera, and some large shells such as the frilled oyster can be found (Fig. 341). Do be careful as these cliffs have fallen on to people. (A few months after we collected there some picnickers were killed.)

Younger rocks of the next stage can be seen on the beach at the mouth of Wairoa Stream, below the dam for the Waipipi Ironsands. The shell beds marking the Waipipian Stage are on each side of the stream, about 100 metres west or 300 metres east, but at high tide the waves wash against the cliffs. Further east, at Waverley Beach, the cliffs are formed of massive mudstone with few fossils and are topped by even younger beach and dune sands (see page 179).

After the Waipipian rocks were deposited the Taranaki area began to rise and the seas shallowed. A thick layer of sand spread over the Waitotara area, forming the creamy cliffs up Waitotara Valley. Some 15 kilometres north of the main road a recent roadcut shows the layers

of shells (especially some large scallops) that define the Waitotaran or Mangapanian Stage (Fig. 342). (The Waitotaran Stage used to be made up of two sub-stages, the Waipipian and the Mangapanian, until geologists made the Waipipian a separate stage. Some geologists prefer to call the other stage the Waitotaran rather than Mangapanian.) Whichever name is used, it denotes the last stage of the Pliocene and the Tertiary Period, the end of the golden weather.

The climate had been changing for a long time. The Oligocene limestones were produced in fairly warm seas and, in the very early Miocene, reef corals (Fig. 343) moved into Northland (page 146), although they have never succeeded in building reefs so far south. However, the seas must have been fairly warm for them to have lived here at all.

Tiostrea lutaria

Fig. 341. Oyster, Ohawe Beach.

Fig. 342. Mangapanian (Waitotaran) layers, just south of the Mangapani Stream, Waitotara Valley.

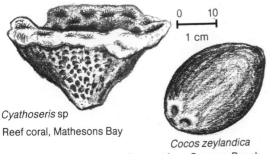

Cyathoseris sp
Reef coral, Mathesons Bay

Cocos zeylandica
Coconut from Coopers Beach.
Note size

Fig. 343. Warm-water fossils in Northland.

The corals were accompanied by other immigrants from the tropics — coconut trees. Tiny fossil coconuts (Fig. 343) and palm pollens have been found at Coopers Beach in Doubtless Bay. There are lignites in the beach banks below the motor camp, and seeds and twigs from the coal measures wash out with the tides.

Since then temperatures have dropped.

THE QUATERNARY PERIOD AND THE COMING OF THE ICE

The Ice Age was not a single long freezing period, but rather a series of colder and warmer intervals known as glacial and interglacial stages. Although the climate had been cooling for some time while ice built up at the poles, about 1.8 million years ago there was a distinct drop in temperature and the ice moved over normally temperate areas.

The spread of ice marks the beginning of the Quaternary Period and the Pleistocene Epoch. In New Zealand the events of these times were recorded in the continuous layers in the Wanganui basin, so the Wanganui Series carries on through most of the Pleistocene.

The glacials affected the coastal waters in two major ways — the seas became much shallower and the animal inhabitants of the seas changed. Sea level throughout the world dropped as the flow of water in many rivers was reduced.

Usually water evaporates from the oceans into clouds, then falls as snow or rain and makes its way down-river to the sea, but during the glacials more moisture remained locked in icefields and glaciers. The seas were not refilled completely. Around New Zealand's coast, areas

where the sea was normally shallow became land, subject to erosion, and deeper waters became shallow. Here mudstones were covered with limestones and sands. In the shallow, colder seas immigrants arrived from sub-Antarctic waters to replace the warm-water animals that had left for warmer waters to the north.

The Antarctic waters today support only a few animal species, but each species exists in great numbers. Similarly, the colder-climate layers of rock generally are filled with a limited variety of animal fossils, and there are some special newcomers. These include a small species of sea-urchin and a deep-water crab that now lives off the Auckland and Campbell Islands as well as around the Otago Coast (Fig. 344).

The most successful of the immigrants was

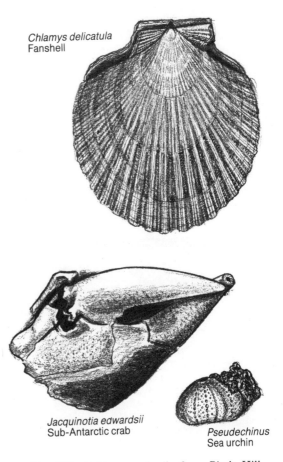

Chlamys delicatula
Fanshell

Jacquinotia edwardsii
Sub-Antarctic crab

Pseudechinus
Sea urchin

Fig. 344. Cold-water fossils from Birch Hill Station, east of Martinborough.

the fanshell *Chlamys delicatula* (Fig. 344), another animal now living off Otago Peninaula. Its arrival in Hawke's Bay is the chief indicator of the coming of the ice. In the Wanganui basin it has been found near Hunterville in the Hautawa Stream, at one end of a shell bed which runs some 32 kilometres across country, drawing a neat boundary line at the base of the next Wanganui Series stage, the Nukumaruan. (The European Pleistocene boundary is now based at 1.63 million years ago, which is high in our Nukumaruan Stage.) Fossils from the Hautawa Shell Bed can be collected on Parapara Road in the cliffs just north of Te Rimu Road, after a scramble up the talus slopes. However, Nukumaruan fossils are widespread and are very easily collected.

Nukumaruan sediments include layers of shallow-water, shelly limestones which are quarried in pits towards Nukumaru Beach from Waitotara. These limestones are covered with sandstones — coarse sediments brought down from the exposed and rising land nearby. Bay bars enclosed wide areas of tidal mudflats with a mixture of muds, sands and shell hash in tidal channels. As the channels shifted across the harbours, layers of debris built out from their sides, producing cross-beds or sloping layers (Fig. 345). These show up very well on the main road west of Kai Iwi at the Ngawaierua Reserve. Here there are beds of shells and sandy layers containing fossil oysters, cockles and also sand dollars — flat, circular sea urchins (Fig. 346).

Other layers of the same age are on the

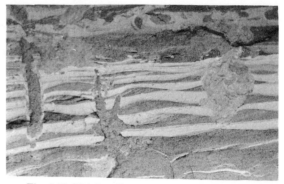

Fig. 347. Tidal sediments and animal burrows, Pipiriki lookout.

Pipiriki road just out of Wanganui, where there is a lookout at the top of the first hill from the turnoff past Upokongaro. Beside the lookout the banks contain shells, and there are also thin, ripple-marked beds of mud and sand which show the tunnels and burrows of animals that lived in these sediments. The tunnels are filled with material from the layers above (Fig. 347).

This entire area was once a shallow tidal flat or coastal plain, where rivers built out deltas and swamps formed. Then the land was crumpled down, and sea covered the swamps with sands and shells. Finally, the rivers built out new deltas and the swamps returned. Fossil pollens in the lignites from these swamps are from plants like those now growing in coastal forests in Southland and indicate a cooler climate once again.

Fig. 345. Cross-beds, west of Kai Iwi.

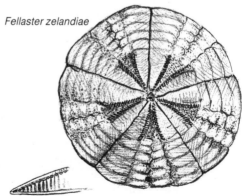

Fellaster zelandiae

Fig. 346. Sand dollar, from the cross-beds.

The effects of the low sea level (caused by the build-up of ice) in the Wanganui area were counteracted by warping down of the crust, one result of the plate movement to the east. The ancient Ruahines were rising, and flats behind the Wanganui area were buckling down; the down-going plate was melting the crust and leading to the first great rhyolite eruptions in the Taupo Volcanic Zone. These events occurred in late Nukumaruan-early Castlecliffian time, so the oldest pumice ashes from the Rotorua-Taupo volcanoes are present in these layers.

As the eruptions continued, pumice sands built up in thick layers which now form cliffs in several places. The Kaimatira Bluffs are located just north of Wanganui on the east side of the river. Here the cross-bedded pumice sands contain shells in their lower parts. There are similar cliffs in the Okehu Stream on the coast road from Maxwell to Mowhanau Beach. In the banks of the road east of the stream are beds of

Fig. 348. Waterworn shells deposited by tidal currents, east of Okehu Stream.

waterworn shells and a few cooler-water fossils (Fig. 348).

Further east down the road at Mowhanau Beach (Fig. 349) grey mudstones provide some of the best and most accessible shell collecting, although again you are warned that the cliffs are dangerous as they are likely to collapse. These mudstones formed as the waters warmed and deepened during one of the interglacial stages. The layers of sandstone, mudstone and shell beds continue right along the coast to Castlecliff and form the type section for the last Wanganui stage, the Castlecliffian.

At Mowhanau, fossil shells may be cut from the cliffs with a knife. They may be cleaned by immersing them in water and gently brushing them free of sand. Some may be found lying down the slope, already washed out by rainwater.

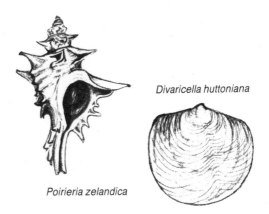

Divaricella huttoniana

Poirieria zelandica

Fig. 349. Cliffs at Mowhanau Beach and fossils found there.

Fig. 350. Upper Castlecliff Shell Bed and some of the fossils from it.

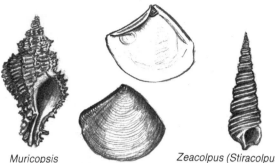

Muricopsis octogonus   *Myadora striata*   *Zeacolpus (Stiracolpus) delli murdochi*

When the tide is right it is interesting to walk the beach from Mowhanau to Castlecliff. Towards the east dunes have built up in front of the cliffs and a sandy path leads over them. Since the cliffs are no longer washed by the sea, grasses grow on the slopes and raupo swamps are ponded between the dunes and the cliffs. Shell beds can still be found. Starting from Castlecliff there are two easily accessible beds within a 20-minute walk from the track at the end of Long Beach Drive.

The shell bed nearest Castlecliff is a sandy layer with a great variety of fossil shells, including warm-water species that arrived from Australia (Fig. 350). The second bed begins further west where the Pinnacles are eroding out on the cliff tops. This bed is a grey, muddy layer containing fossil oysters, bryozoans and brachiopods and the first appearance of the scallop, *Pecten tainui* (Fig. 351).

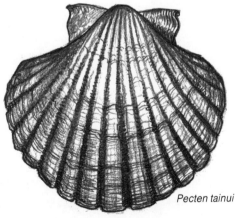

Pecten tainui

Fig. 351. Tainui Shell Bed, Castlecliff, and fossil scallop.

Hawera terraces — upper Quaternary

So far we have looked at the sloping layers that fill the Wanganui basin, but along the tops of the cliffs there are also level layers, giving the Wanganui area the appearance of a series of giant steps (Fig. 352). The change from sloping layers to level tops marks the transition to the final series of rocks made by the alternating changes of glacial and interglacial climates.

In Europe four main ice advances have been named; but here some six have been described, roughly corresponding to the two Wanganui sub-stages (Hautawan and Okehuan) and the four main terraces described in the Hawera Series.

Terraces were cut all around New Zealand at times of high sea levels in the interglacials. The absolute height of those sea levels in relation to current sea level is very hard to estimate because the heights of many terraces have been altered by earth movements, which were different in each area.

The Taranaki-Wanganui terraces are cut on to the Wanganui layers and seem to represent an almost uninterrupted series of events. The

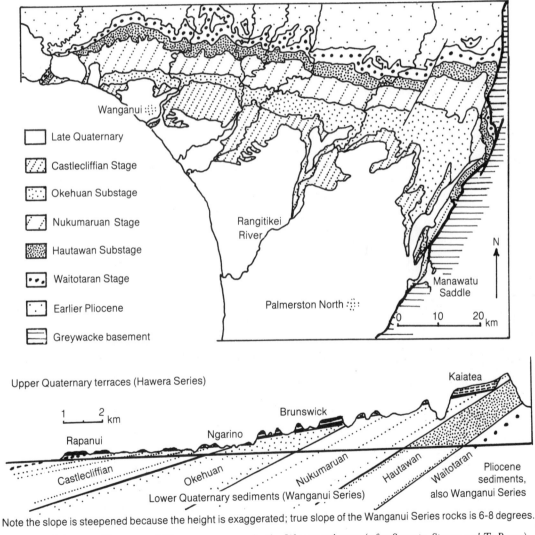

Late Quaternary

Castlecliffian Stage

Okehuan Substage

Nukumaruan Stage

Hautawan Substage

Waitotaran Stage

Earlier Pliocene

Greywacke basement

Wanganui

Rangitikei River

Palmerston North

Manawatu Saddle

N

0   10   20 km

Upper Quaternary terraces (Hawera Series)

1   2 km

Kaiatea

Brunswick

Ngarino

Rapanui

Castlecliffian

Okehuan

Nukumaruan

Hautawan

Waitotaran

Pliocene sediments, also Wanganui Series

Lower Quaternary sediments (Wanganui Series)

Note the slope is steepened because the height is exaggerated; true slope of the Wanganui Series rocks is 6-8 degrees.

Fig. 352. Wanganui layers and Hawera terraces in the Wanganui area (*after Suggate, Stevens and Te Punga*).

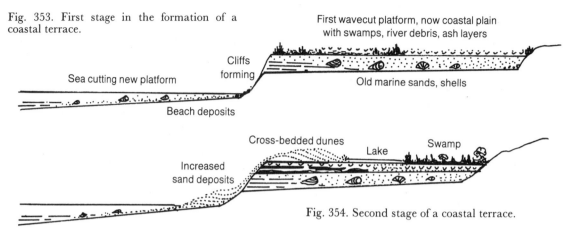

Fig. 353. First stage in the formation of a coastal terrace.

First wavecut platform, now coastal plain with swamps, river debris, ash layers

Cliffs forming

Sea cutting new platform

Old marine sands, shells

Beach deposits

Cross-bedded dunes

Lake

Swamp

Increased sand deposits

Fig. 354. Second stage of a coastal terrace.

slope of the Castlecliff layers and the top Hawera terrace is almost the same, but the top terrace is some 15 kilometres inland and 150 metres higher. The angles of all the other layers are different, showing the effects of continual tilting — the older the rock the more it has been tilted.

The sea seems here to have moved inland without any earth movement and cut a cliff at the back of a wide tidal platform. The land beyond the cliff was of soft Wanganui Series rocks which washed into the rivers. East of

Wanganui the rivers brought down sufficient of this debris to build out bulwarks against the sea, which broke against this new material instead of cutting cliffs; the terraces were not continuous. Then sea level dropped and the sea began to cut a new, lower platform while the old sea bed became a coastal plain bearing grasses and river beds (Fig. 353).

About this time the volcanoes in Taranaki and those of the central plateau were erupting, producing ash and the first ironsands found in the coastal deposits. On the beaches sand dunes

Fig. 357. Sand dunes showing cross-bedding, Awakino.

178

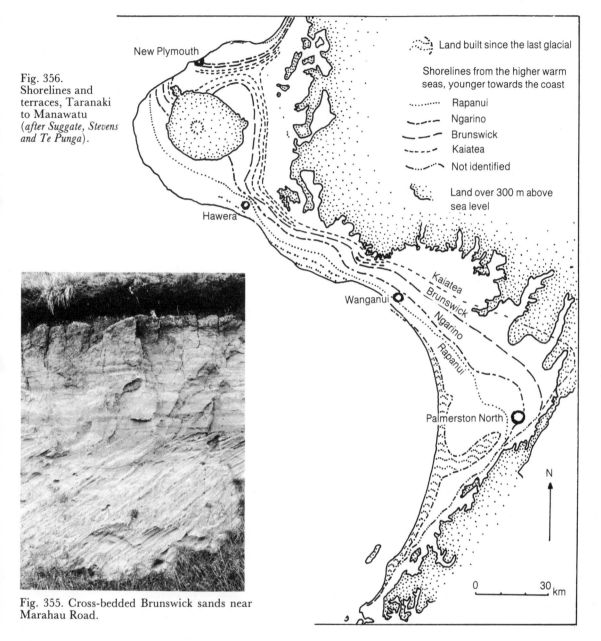

Fig. 356.
Shorelines and
terraces, Taranaki
to Manawatu
(*after Suggate, Stevens
and Te Punga*).

Fig. 355. Cross-bedded Brunswick sands near
Marahau Road.

built up then moved inland across the former
platform as they do today. The dunes dammed
streams to form little lakes and swamps (Fig.
354). As sand moved up the dunes and flowed
down their backs it built up in cross-bedded
layers, like those in the roadcut 200 metres east
of Marahau Road, just west of Birch Park, in the
sands of Brunswick Terrace (Fig. 355). The
same processes were repeated until all the

current terraces were formed (Fig. 356).

Along the sea cliffs today layers of the lowest
Rapanui Terrace are exposed. At Waipipi
marine sands and muds are draped over boulders of the older shell bed. At the top of the cliff
the old layers of the sea bed are covered with
swamp deposits. Some of the most spectacular
dune layers are visible at Awakino alongside the
main road (Fig. 357).

179

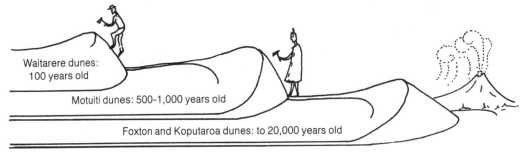

Fig. 358. Four sets of dunes in the Manawatu.

Currently dunes are building up and moving inland as part of a continuing series. Older dunes in the Rangitikei-Manawatu area have stabilised, and soils have formed on them; nearer the coast younger rows of dunes are made of fresher sands. The growth of dunes has not been constant; they have advanced with increases in the sand supply carried by the rivers.

In the Manawatu there are four sets of dunes (Fig. 358). Beneath the oldest dunes is Aokautere Ash from an eruption at Taupo 20,000 years ago. Possibly ash fall and associated fires removed the plant cover from the land, increasing erosion and the cold climate slowed their regrowth. More recently, Maori and pakeha people have cleared the forest and renewed erosion.

On the banks of the Manawatu River between the State Highway 1 bridge and the Whirikino Cut downstream several layers of dunes can be seen, each indicating a period of dune advance after renewed erosion. We know that all these have built up in the last 2,000 years because Taupo Pumice from the eruption of A.D. 186 is present at the base of these dunes.

When the supply of sand decreased in the Wanganui area the wind blew the dunes inland, exposing the older layers. Near Nukumaru Beach the younger sands are gone, except for rusty remnants, and wind-blasted limestone forms hummocks in the lupin and scrub. Here also pebbles from the old beaches have remained and have been wind-blasted into new shapes, like brazil nuts or Chinese hats (Fig. 359). These ventifacts (meaning "wind-mades") are protected at Waitotara because they are no longer being formed. However, they can be found in the lag gravels among the dunes north of Deadman Stream above Clarence River in Marlborough, where the white Amuri limestone has taken on unusual patterns (Fig. 360). Ventifacts are also present on the beaches at Awarua Bay, on the estuary north of the aluminium smelter at Tiwai Point.

Fig. 359. Shapes of ventifacts.

Fig. 360. Marlborough ventifacts, Deadman Stream.

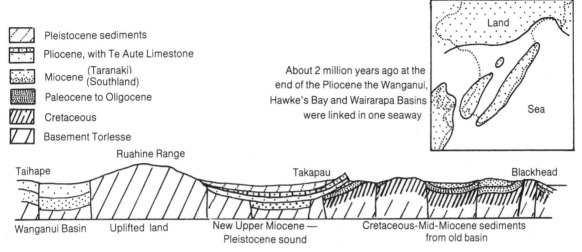

- Pleistocene sediments
- Pliocene, with Te Aute Limestone
- Miocene (Taranaki) (Southland)
- Paleocene to Oligocene
- Cretaceous
- Basement Torlesse

About 2 million years ago at the end of the Pliocene the Wanganui, Hawke's Bay and Wairarapa Basins were linked in one seaway

Land

Sea

Ruahine Range

Taihape   Takapau   Blackhead

Wanganui Basin   Uplifted land   New Upper Miocene — Pleistocene sound   Cretaceous-Mid-Miocene sediments from old basin

Fig. 361. Pattern of land and sediments in the Southern North Island.

The east coast

While the Wanganui basin was filling there lay another basin down the east coast from Gisborne to Palliser Bay. The old eastern sea from Hastings south was squeezed up at the end of the Miocene and a new seaway or sound formed between the two main ranges and the new coastal ridge (Fig. 361). The cross-section shows where layers of sediment built up as the seas spread further onto the ranges, crossing them at the Manawatu Gorge and the Taruarau Saddle, by Kuripapango. Out to the east a series of squeezed-up humps with greywacke cores and older Tertiary topping enclosed the new sea channel.

The shallow sea was floored with sandy mudstones and shellfish-bearing sandstone beds in which concretions formed. This area was constantly pressured by the moving plate boundary; there are small breaks in the rock sequences and the sea varied in depth from place to place. One of the few areas where a long sequence of Tertiary sediments can be seen is the valley behind Te Mata Peak at Havelock North, where a narrow sound persisted in the northern end of the old seaway. Here the whole Tertiary sequence from Eocene to Pliocene has been crumpled up and cut open by the Tukituki River (Fig. 362, p. 182).

The blue-grey to white mudstone in the river just above the Waimarama bridge is Eocene in age with no large fossils and very few forams. Down-river to the north the banks are of Oligocene mudstone, which by Joll Creek con-

tains many large and visible forams. The main road below Te Mata Peak runs through massive pale grey Miocene mudstone which is fracturing as it dries out and which contains a few fragile shells. Where the uppermost Miocene rocks should be found the rocks are covered with material slumped off the peak; it is Pliocene sandstone at the base and upper Pliocene limestone at the top — Waitotaran Te Aute Limestone.

There is a walkway up the cliff with good views from the top. The layers are repeated to the east of the river. From this viewpoint you can see the Eocene mudstones in the river, the Miocene mudstones and the limestone hilltop behind (Fig. 363). In the quarry in the middle

Fig. 363. Looking east across the Tukituki River.

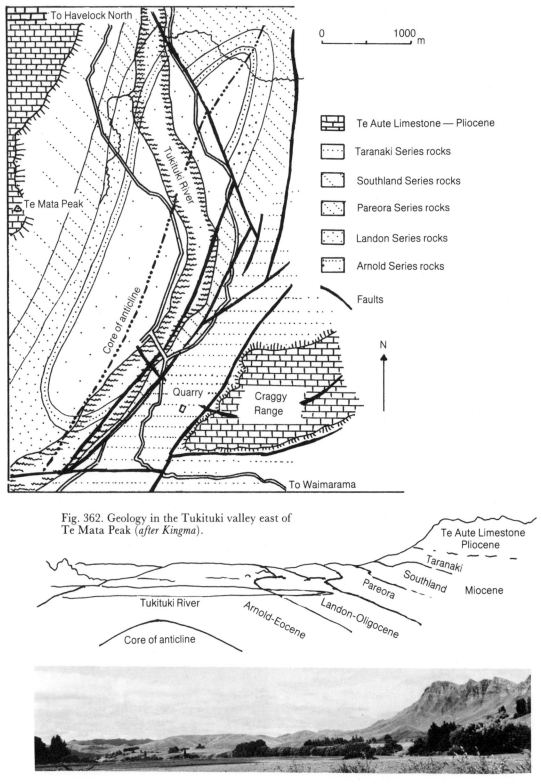

Fig. 362. Geology in the Tukituki valley east of
Te Mata Peak (*after Kingma*).

distance redeposited limestone is worked.

The central Craggy Range Stream carried lime dissolved from the Waitotaran Limestone on top of the range down a gorge cut by the stream into the Miocene mudstones. The lime was deposited in thick layers of brown travertine and contains leaf impressions and rare fossil moa bones. This deposit formed in the last few thousand years. It is of the same composition as the layers that build up inside water pipes and kettles in limestone country.

Te Aute Limestone extends in great ridges all through central and southern Hawke's Bay. It is the material quarried for fertiliser at Pakipaki, Waipawa and Hatuma. It is composed of broken shell fragments cemented together — especially barnacle shells, which are a feature of these young limestones. Whole shells present in the rock are generally oysters, fanshells and large cockles (Fig 364), while there are also some bryozoans and sharks' teeth. The Hatuma Quarry is sited across a bed where fossil teeth were fairly common; examples are displayed in their office. However, such fossils are more difficult to find now that power machinery is used for quarrying.

The Te Aute Limestones formed when seas were shallow over the area and strong currents washed finer silts and muds from the drifts of shells. The shell banks shifted in time, causing the limestones to be of different ages across the seaway (Fig. 365).

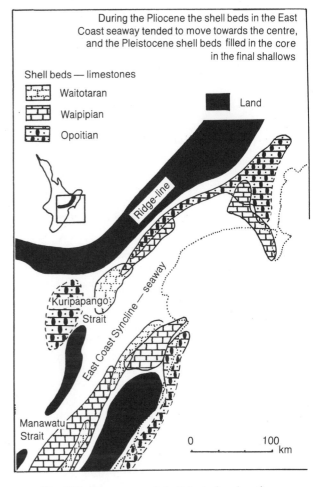

During the Pliocene the shell beds in the East Coast seaway tended to move towards the centre, and the Pleistocene shell beds filled in the core in the final shallows

Shell beds — limestones

Waitotaran

Waipipian

Opoitian

Land

Ridge-line

Kuripapango Strait

East Coast Syncline — seaway

Manawatu Strait

0   100 km

Fig. 365. Movement of shell beds forming the limestones in the Hawke's Bay area (*after Beu et. al.*).

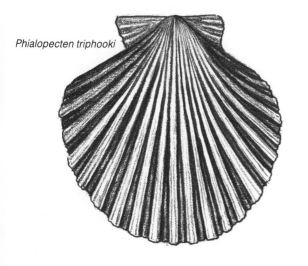

Phialopecten triphooki

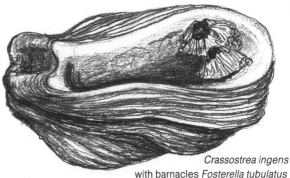

Crassostrea ingens with barnacles *Fosterella tubulatus*

Fig 364. Waitotaran fossils in the Te Aute Limestone.

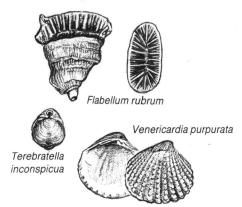

Flabellum rubrum

Venericardia purpurata

Terebratella inconspicua

Fig. 366. Nukumaruan fossils from Maraekakaho.

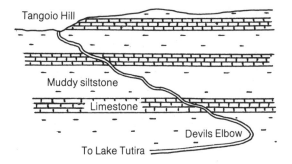

Fig. 367. Layers of muddy siltstone and limestone on the north of Tangoio Hill.

In the cold Pleistocene seas the same pattern persisted; sands and shell beds drifted with the currents across the shallow sea floor.

Near Maraekakaho, where State Highway 50 passes Whakapirau Road, sandy banks contain small brachiopods and little solitary corals (Fig. 366). In other areas layers of shells have been cemented together into a hard limestone, blue-grey in colour. This rock differs from the Te Aute layers in that the large shells are usually absent; they were dissolved away as their lime provided the cement for the rock. In some cases the spaces left by the shells were filled with calcite crystals.

Nukumaruan sandy limestone forms the cliffs of Napier and is found in the hills west of the plains almost as far north as Wairoa. Here, however, the hillsides are composed of layers of limestones alternating with muddy siltstones which were laid down as the seas deepened in the warm interglacial periods. Judging by exposures in the western hills, there could be eight or nine sandy shoreline layers which formed in the colder periods.

On State Highway 2 north of Tangoio Hill, from Kaiwaka Road down towards Lake Tutira (the Devils Elbow), layers alternate in the roadcuts (Fig. 367), and there are cross-beds in the sandy limestones and scattered shells in the muddy siltstones. These layers are also visible on the hills above Lake Tutira.

Further north at the site of Matahorua Viaduct the water had deepened. Layers of sand and silty mudstone here show many feeding burrows and trace fossils, but few shells (Fig. 368).

In the south limestones cap the hills near Dannevirke. In the Wairarapa younger layers are the main source of agricultural lime from that area.

Fig. 368. Deeper-water sediments by Matahorua Viaduct.

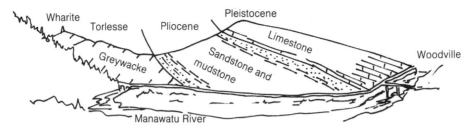

Fig. 369. The east end of the Manawatu Gorge.

In the Manawatu Gorge the view north from the hill above Ballance Reserve shows Pliocene and Pleistocene layers sloping up the hump of the greywacke range (Fig. 369). Nukumaruan limestones on top form the scarp and quarry over the railway tunnel. Underneath are older grits, sandstones and siltstones, and on the far left is the greywacke of Wharite Peak.

Nukumaruan grey siltstones occur just south at Marima Domain. Here the banks of the Mangahao River are studded with shell fossils. Similar grey siltstones form the banks at Popes Head, east of Martinborough, where fossils are weathering out of a massive roadcut (Fig. 370). Further up the hill the composition of the banks changes to a sandy, brown limestone containing shells, some of which are filled with crystals of calcite.

The effects of glacial climate changes appear also at Castlepoint on the Wairarapa Coast (Fig. 371). The reefs and castle here are composed of a sliver of rock much younger than the landward Miocene rock discussed in the section on Whakataki (page 166); the reef is Nukumaruan in age. Some of the Castlepoint layers of sands and siltstone are dominated by fossils of the cold-water *Chlamys delicatula*, while others contain a rich variety of warmer-water shellfish remains (Fig. 372).

Be wary when exploring the reefs at Castlepoint because the waves break over the reef much higher than you would expect. With care

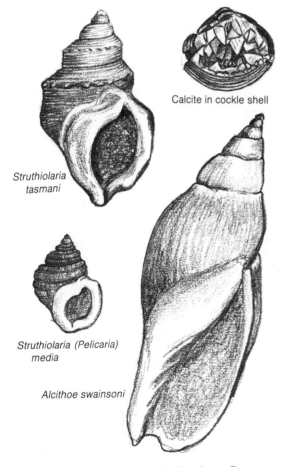

Calcite in cockle shell

Struthiolaria tasmani

Struthiolaria (Pelicaria) media

Alcithoe swainsoni

Fig. 370. Nukumaruan fossils from Popes Head, east of Martinborough.

Fig. 371. The castle and reefs, Castlepoint.

you can find on the path up to the lighthouse and on nearby ledges places where shells are weathering out of the sandy banks. Up on the castle the fossils present include many bryozoans as well as fanshells and brachiopods, a few

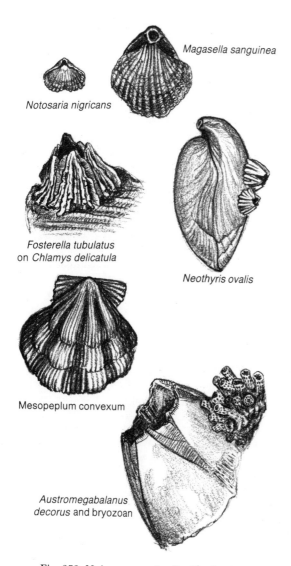

Magasella sanguinea

Notosaria nigricans

Fosterella tubulatus
on *Chlamys delicatula*

Neothyris ovalis

Mesopeplum convexum

Austromegabalanus decorus and bryozoan

Fig. 372. Nukumaruan fossils, Castlepoint.

of which are hollow and show the delicate loop of shell that supported the feeding fringe of the brachiopod.

By the end of the Nukumaruan Stage continuing pressure along the east coast caused by plate movement was raising the seaway, folding the rock layers so that gradually all the sound became a very shallow sea and then river plain, with freshwater muds and gravels covering the sea-floor sands between the lifting ranges. As the pressure increased, the hard, old basement rocks cracked and broke, forming fault cliffs or scarps.

Looking south near Takapau freezing works on State Highway 2 you can see humps of limestone that have been lifted along fault lines. (Fig. 373). They rise above the old sea basin which became a river plain in the same way that the shallow lagoon on the west side of Napier was lifted into a plain in the 1931 earthquake.

The transition from marine sands to river gravels shows up especially well in the cliffs between Clifton and Cape Kidnappers (Fig. 374). The Lands and Survey Department has published a good guide to the geology of the area for the Cape Kidnappers Bird Sanctuary Board.

As you walk along the beach you take a journey back in time. At Clifton the rocks are about 300,000 years old, formed at the end of Castlecliffian time. Imagine the landscape as it was then, as it is now further inland — a vast, shallow bay with some tidal areas where cockles grew but where the main Hawke's Bay rivers were also building up a flood plain of gravels, boulders and peat swamps, covered from time to time with thick layers of pumice from eruptions in the central volcanic area. The past record of this plain is preserved in the cliffs all the way to Black Reef, where the oldest layer is a shell bed from an ocean shore. The shellfish are those species present on a sandy beach today — tuatua, oyster, cockle and whelk — but here they are a million years old.

Beneath this beach is the hard limestone of

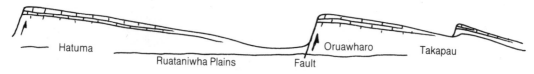

Hatuma

Ruataniwha Plains

Fault

Oruawharo

Takapau

Fig. 373. Fault scarps near Takapau.

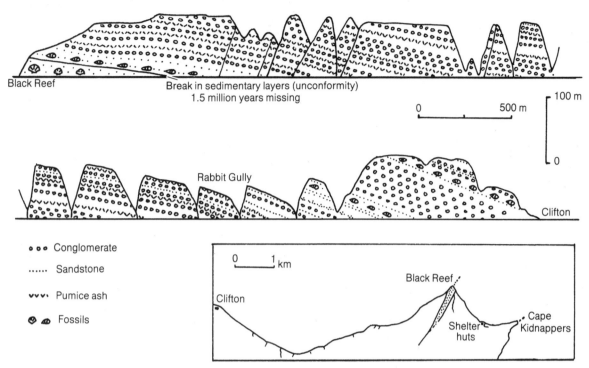

Fig. 374. Sediments on the way to Cape Kidnappers (*after Kingma*).

Black Reef itself, containing layers of fanshells about two and a half million years old. What happened in the missing one million years?

Consider the inland sea channel running to the south and west and imagine currents sweeping up past the Te Aute ridges, carrying away any sediments that might have settled in the area. Look at the Black Reef rocks and picture them as a sea bed swept clean of muds and silts by currents which enabled a shell bottom to collect in relatively shallow waters.

Then as you walk towards the cape see how the sediments in the cliff change in colour and texture. At first there is grey sandstone scattered with rather fragile fossils of shells now found at a depth of 20 to 50 metres; you have moved back to a time when the seas were deeper. By the shelter huts you are five million years back, on the deep sea floor where the only shells that fell to rest were a few forebears of the oceanic violet snails that wash up today on our beaches. These rocks take you back to the beginning of the Pliocene, the beginning of Wanganui time when this area was open sea.

Look up at the flat terraces on top of the cliffs. They will bring you forward in time again to the last 100,000 years, to the glacial and interglacial changes of sea level.

Pliocene-Pleistocene rocks in Canterbury

Flat terraces on top of sea cliffs are a feature of the Canterbury coast at Motunau, where underneath the terrace deposit we can see the same deep-water grey siltstones of Pliocene age as those at Cape Kidnappers.

When we first looked at the layers along the Waipara River (page 112) we discovered how the Cretaceous land had gradually been covered by the sea. In Canterbury an identical blanket of coal measures, greensands and limestone layers stretched from the Waitaki Valley, past the Cretaceous volcanics at Mount Somers and the Rakaia Gorge and far inland as well as to the north.

At Porters Pass the limestones in the Castle Hill basin seem now to be cut off from the sea,

187

Fig. 375. Limestone and greywacke
at Castle Hill.

but during the Oligocene Epoch the mountains were still down in the basement underneath the shallow sea.

Travellers who turn aside at Cave Stream Scenic Reserve, 100 metres west of Broken River Bridge (the turnoff is on a bend and rather hard to see), may enter a limestone cave where the stream goes underground. It is possible, using swimsuits and waterproof torches, to go all the way upstream through the cave, although anyone looking down the hole where the water plunges from the upstream cave may doubt the passage exists.

Towards Castle Hill Station the limestone layers tilt steeply up to the Torlesse Range (Fig. 375). The forces that raised the Kaikouras and Southern Alps and were twisting the Wairarapa and Hawke's Bay hills in the Pliocene also raised the block of greywacke from the basement to form a towering range; and while the brittle, older greywacke sandstones broke along the faults each side of the range, the softer and more pliant younger rocks bent under the same pressure.

Similar folds in the rocks of the Tertiary can be seen in the North Canterbury hills where cores of greywacke hold up younger rocks, and rivers and roads tend to follow the synclines between.

In the upper Waipara (page 112) we looked at the shelly Mount Brown limestone capping the rock layers, indicating a shallow Miocene sea in the west. At the same time in the east deeper

seas were floored with thick, grey silts.

Going east at the Waipara intersection where the Lewis Pass road meets State Highway 1 you can cross the Cass anticline where the Tertiary layers are visible in the hills (Fig. 376), although the road turns north and follows the Miocene Southland beds to the coast at Glenafric Station.

The cliffs each side of Dovedale Stream contain fossils which weather out on the beach (Fig. 377). Most notable are fossil crabs with giant claws, and nautiloids, both present in concretions and not easy to find. In some cases fossil crabs have been eroded from softer material so only a remnant of the body is found; in others the concretion is so hard that only a small part of the crab is visible (breaking such a concretion usually results in a broken crab).

Concretions here also contain shells, whalebone and a species of white, fan-shaped coral, and limestone boulders contain shells and sharks' teeth. Also weathering out of the cliffs and collecting in gravelly areas and among tidal pools are nodules of pyrite and pyritised casts of fossils, some very rusty and others fresh enough to shine after a bath in acid.

The Canterbury siltstones continued to collect on the sea floor into the Pliocene. The younger Pliocene beds are visible in the Greta Valley and at Motunau, where late Pliocene crabs have been collected.

On the way to Motunau the road crosses the familiar series of slumping Eocene mudstones and a scarp of white limestone, but beyond these

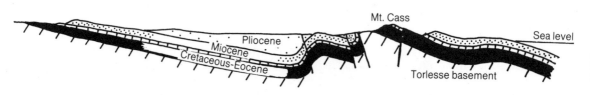

Fig. 376. Folded layers in North Canterbury at Waipara.

Fig. 377. The mouth of Dovedale Stream at Glenafric Station and fossils found on the beach, of Southland age.

Flabellum pavoninum in concretion

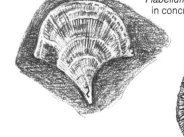

and pyritised.

Pyritised brachiopod,

coral

and gastropod.

Pyritised nautiloid; much larger nautiloids may be found with the original shell

Tumidocarcinus giganteus

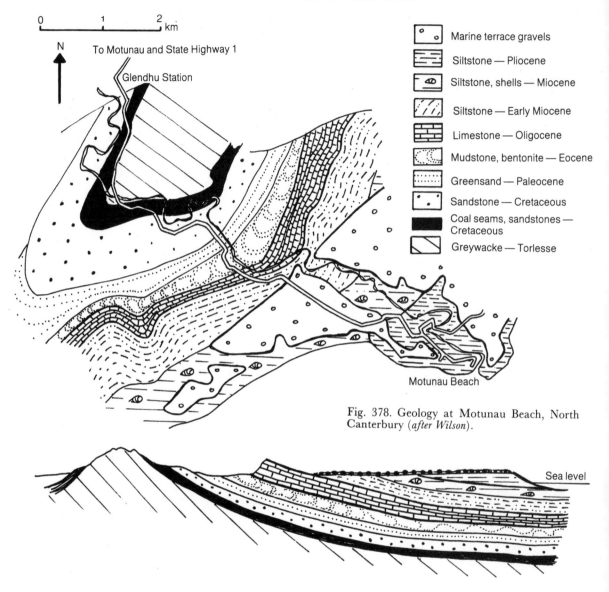

Fig. 378. Geology at Motunau Beach, North Canterbury (*after Wilson*).

the grey Miocene-Pliocene sediments continue out to sea (Fig. 378).

South of Motunau the cliffs on the beach have yielded gypsum clusters, and from the reefs and shore boulders have come rare fossil penguin bones and concretions containing a frilled crab and a small crayfish species (Fig. 379). There are also silicified algal nodules and whalebone to be found when the tide is right.

Further inland many of the Pliocene deposits contain gravels and conglomerates from the rising land. These show up very well on the roadcuts on State Highway 1 in the Hundalee Hills north of the Conway River, where the road passes through thick marine gravels and sandstones as well as the underlying Torlesse rocks.

The transition from shallow marine sands to freshwater gravels is visible up the Leader Valley and marks the final emergence from the water of the deep-sea slope that once bordered our land from East Cape to Canterbury.

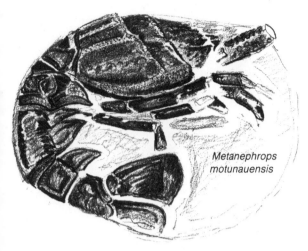

Fig. 379. Pliocene fossils from Motunau Beach.

Metanephrops motunauensis

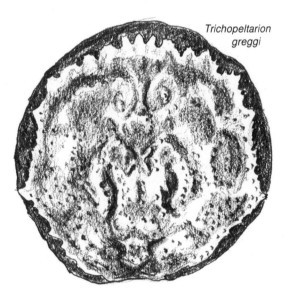

Trichopeltarion greggi

Ice Ages may have come when the drifting plates moved the land mass away from the pole. Cycles of warmer and cooler periods may be linked also to changes in the direction and amount of tilting of the Earth and its orbit around the sun. In the last Ice Age Antarctica may have remained more or less frozen while New Zealand responded to the world's climatic pendulum.

During the glacials, ice built up in the form of glaciers and ice caps. It moved across the land, gouging out deep, basin-like hollows and scoring the rock surfaces, using pieces of rock frozen into the base of glaciers like blades in a carpenter's plane.

Glaciers leave a characteristic wide-bottomed, steep-sided valley with smaller basin-like valleys on the sides where the tributary glaciers joined the main glacier (Fig. 380). These are the typical valleys of Fiordland and the upper reaches of the South Island rivers. A few valleys of this shape can also be found in the Tararua Ranges.

Most of the large lakes on the east of the Southern Alps lie in hollows carved out by glaciers. The ice-smoothed sides of the mountains often contrast with the jagged peaks above, showing the depth of the original ice. For example, at Lake Wakatipu just one mountain, Nicholas, has a rounded head showing where an arm of the glacier stretched out towards Oreti Valley.

Another arm of the Wakatipu Glacier once extended over to the Arrow Basin, joining the Shotover and Arrow Glaciers to flow down Kawarau Gorge. It rounded View Hill and all the peaks within the basin, forming them into

The glaciers

So far we have looked at some of the effects of the Ice Age on the marine sediments and coastal wavecut terraces. Ice left its mark on the South Island as well, chiselling out the uplands and depositing special patterns of sediments as the glaciers alternately advanced and retreated.

The causes of the great period of cooling are still not perfectly clear. There were earlier Ice Ages in the Ordovician and the Permian, when each time the polar regions were covered with large land masses where ice-sheets built up as they do now in Antarctica. The end of the past

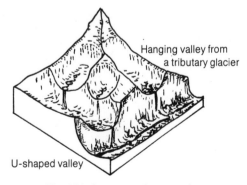

Hanging valley from a tributary glacier

U-shaped valley

Fig. 380. Ice-scoured mountains.

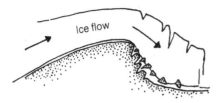

Fig. 381. Roche moutonnée.

roches moutonnées (sheep-shaped rocks) — isolated, rounded hills with a smooth side where the ice rode up and a rougher, steeper side where the passing ice plucked out pieces of rock (Fig. 381). Lake Hayes lies in the glacially scoured basin, surrounded by glacial features.

Fiordland was also shaped by ice. Here the glaciers gouged out their paths to the sea; the sheer walls of the valleys show the great depth of the original ice, and the smaller side valleys now contain little lakes in their basin-like floors and some spectacular waterfalls where streams plunge down into main valleys. Peaks at the heads of these valleys were cut back as the ice formed circular basins or cirques where the glacier began. Mount Aspiring is typical of the horn-shaped peaks between glaciers.

Frost-shattered pieces of rock which fell from cliffs above the glaciers were carried along on top of the ice as moraine, just like the rock debris that now buries the Tasman Glacier for much of its length. Rock fragments carried by ice are not sorted or rounded like those carried by water, so when the ice melts they are dumped in vast piles of rocks of all shapes and sizes, from finest dust

to enormous blocks. Even when erosion starts to sort the debris there are rocks too large for water to move and these are left standing out as glacial erratics, such as those that stud the Tourist Gardens in Queenstown. The piles of moraine from old glaciers extend far down South Island valleys, showing the original extent of the ice. They form ridges south of Frankton; they pond lakes such as Clearwater in the east and Matheson in the west; and they form the forested, snakelike hills between the rivers on the west coast.

Where a large chunk of ice remains in a moraine it melts to make a hole in which forms a kettle lake, like the pond where the tourists stand for a view of Fox Glacier (Plate 7E). This scene also shows the steep front of the uplift that occurred along the Alpine Fault. (In the visitors' centres at the glaciers there are very good displays on the geology of each area and the effects of the ice.)

Although ice never smoothes the rocks it carries, it does grind those in its bed to an extremely fine clay or rock flour, which colours the waters of glacially fed rivers milky blue.

While the uplands were covered in ice the land further out supported scrub and tussock as the forests retreated before the cold (Fig. 382). Blowing across the treeless wastes, gale-force winds picked up the silts and rock flour from river flats and carried it far across the land, dropping it eventually to be trapped in the grasses and to form a mantle of loess, recognisable in roadcuts by its fluted, columnar weathering (Fig. 383).

Fig. 383. Loess on Banks Peninsula.

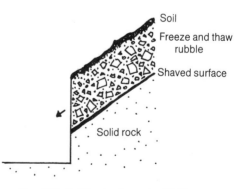

Fig. 384. Solifluxion layer in Wellington.

In South Canterbury loess lies 15 metres deep over the basalts that flowed down to the sea at Timaru from Mount Horrible at the beginning of the Ice Age, when a smaller basalt flow also banked up behind Geraldine (Plate 7F).

The cold winds that lifted the loess also blew sand against the pebbles and stones left on the plains, producing more ventifacts. In Central Otago large chunks of cemented quartz grits were wind blasted and polished where they lay eroded on the slopes, especially in Ida Valley (see Fig. 264, p.135) and on Raggedy Range.

In some areas away from the ice the ground was frozen at depth, but the top layers thawed out in warmer weather. Rain or melted frost turned the upper layers of shattered rock to a slush which moved down hillside slopes, shaving a surface across the frozen rock below. In many Wellington hillsides and roadcuts such soil flows or solifluxion debris can be seen. When the foot of the slope is cut away in subdivisions these flows can be unstable — after rain the waterlogged rubble once again slides down the shaved, smooth surface between it and the solid rock below (Fig. 384).

Fig. 382. New Zealand in the last glaciation, about 20,000 years ago (*after Suggate, Stevens and Te Punga*).

Area above snowline

Shoreline

Outwash gravels

Warm-climate forest

Cool-climate forest

Tussock and scrub

Alpine plants

Volcanoes

Rhyolite

Andesite

Basalt

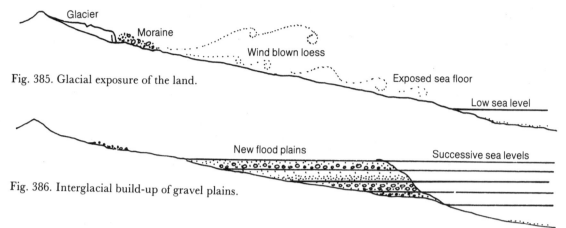

Fig. 385. Glacial exposure of the land.

Fig. 386. Interglacial build-up of gravel plains.

In contrast, rocks in other areas of Wellington weathered into a rich, red soil like the red earth found in North Auckland and Coromandel Peninsula. The red colour in the soil is hematite, oxidised iron which forms in soils in warm climates. It would seem to have formed in Wellington in the warmer periods between glacials and is obvious in the roadcut by the badminton hall at Hataitai.

Meanwhile, every river valley in the country was cut and filled to form a new shape by the alternation of glacial and interglacial climates. As the amount of water in the ice cap increased during a glacial period sea level dropped and around New Zealand a wide expanse of old sea floor became bare and windswept new land. Here winds added sea floor silts to the glacial dust forming the loess. Rivers cut deep across lowlands as they worked their way down to new, lower sea levels. See Fig. 385.

Then as the glaciers retreated in a warming period, vast amounts of rock debris were re-

leased from the frost and added to the loads carried by rivers. Until the land could be mantled in protective forests, falling rain washed vast quantities of debris down water-courses onto the plains — pebbles and boulders of all sizes mixed with fine silt and known as outwash gravels.

As the seas rose rivers built their flood plains higher to match the rising sea level (Fig. 386). Vast plains of outwash gravels built up in Canterbury and, to a lesser extent, on the strip of the West Coast and across the Moutere Valley where the hills are now dissected by rivers but where roadcuts show deposits of old, weathered boulders (Plate 7B, C).

With the succeeding glacial period rivers again cut deep into boulder plains, and while this was happening a blanket of loess was laid on terrace surfaces (Fig. 387). In the next warm period the new river beds were filled. The cycle continued, building up the flights of terraces which can be seen in nearly every major river valley around the country, even those without

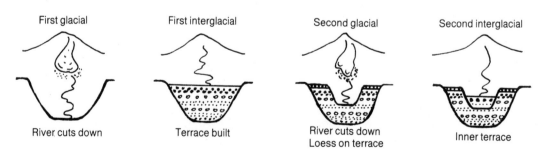

Fig. 387. Formation of flights of river terraces with loess.

glaciers in their headwaters. See Fig. 170, p.90.

With the variations of climate some of the older species of plants and animals were lost and others were confined to small areas where separate species developed. Other creatures arrived as warmer currents moved down again from the north, and still more drifted in on the cooler seas that circle the oceans between Australia and Antarctica.

The development of our contemporary plant and animal communities in the 10,000 years since the ice retreated is properly dealt with by biologists, and the final shaping of our landscape by ice and running water is a subject that has been covered very well by other writers whose books are listed on page 218. However, although we are almost at the end of the sedimentary story we still have to consider the most recent volcanic rocks and the movements that are lifting and stretching our land in the continuation of the Kaikoura Orogeny.

The effects of earthquakes

Looking again at the current geological state of New Zealand (Fig. 388) we can see the relationship between the plate boundary, the volcanoes and the movement of the land.

All the grey Miocene sediments, especially the Great Marlborough Conglomerate (page 160), record the beginning of the uplift some 15 to 20 million years ago as the first large blocks of our land were rising from the sea. Since that time movement has been almost continuous as the two plates crunch together. In older areas the faults and boundaries made in the Rangitata collision controlled the movement, causing the younger Tertiary sediments to fold on top, as we saw near Murchison where the Miocene sediments stand upright in an area where movement has been somewhat notorious. Along the east coast seaway in the North Island a newer series of faults now causes disruption of the land.

Because the plate boundary lies at an angle, movement occurs in two directions — the land is both compressed and squeezed upwards and there is also sideways slippage. Most of New Zealand's recent faults show both vertical and lateral displacement.

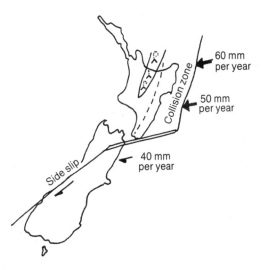

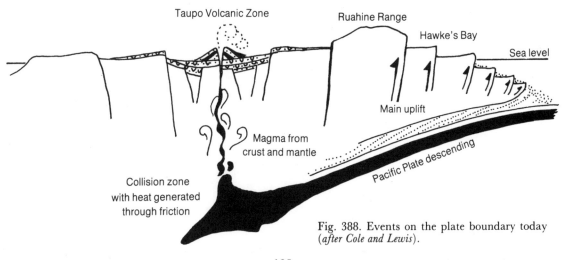

Fig. 388. Events on the plate boundary today (*after Cole and Lewis*).

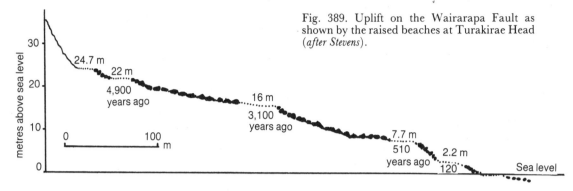

Fig. 389. Uplift on the Wairarapa Fault as shown by the raised beaches at Turakirae Head (*after Stevens*).

The recent uplifts are familiar to most of us. The airport and plains around Napier rose from the sea in 1931, and the 1855 Wellington earthquake raised beaches to make a platform for the Marine Drive and Hutt Road. The proposed sheltered anchorage for ships became the Basin Reserve playing field. At Turakirae Head (Fig. 389) between Wellington and Palliser Bay a series of raised beaches show that fault movement has been continuous and the rate fairly even for the last few thousand years.

Contemporary uplift has been estimated at about 13 millimetres per year. If that rate has been constant, within a million years the land would have been lifted some 13 kilometres; over two million years the rocks would have been raised some 26 kilometres. Since this is about the depth at which the Haast Schist could have recrystallised from greywacke we know that this amount of uplift has occurred in the past, and it seems it could have taken place in the last two million years or more recently. (The steep front on the west of the Southern Alps is further evidence of a rapid and recent uplift.) However, not everyone agrees the rate of uplift has been constant for so long. Some geologists believe there was less movement in the past or that the

movement occurred in fits and starts. Six million years' uplift is one estimate.

There is even less agreement about the amount of lateral movement during the last few million years. At present the faults in Marlborough, Nelson and southern North Island are showing evidence of considerable sideways movement. At Branch River, up the Wairau Valley, river terraces have been repeatedly offset, and older terraces show more movement than those formed later (Fig. 390). You may walk along the fault line here and see where the movement occurred. There is room to park a car by the river, and the farmer requests that walkers be careful of his land.

Similar movement has displaced terraces on the Waiohine River in the Wairarapa and the Hutt River at Harcourt Park, Upper Hutt. In the Buller Gorge there is a signpost at White Creek, showing the uplift of the fault visible across the river which occurred as a result of the 1929 Murchison earthquake. In the bush above the main road a water race was offset both vertically and laterally.

Some geologists would argue that the same amount of movement as today, extended over the last 3 million years would have put the rocks

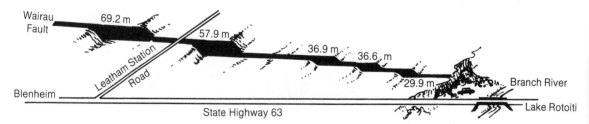

Fig. 390. Movement on the Wairau Fault shown by the offset of the Branch River terraces (*after Stevens*).

of Nelson and Otago together in the Pliocene, but many others believe the great South Island displacement happened much earlier. Furthermore, we are unsure whether the rocks of the North Island were displaced in the same way and over a similar distance. If this were the case we would wish to know whether the movement occurred along only some of the many fault lines or all of them (Fig. 391).

Although the latest movements are well documented, many unanswered questions remain about the effects of past earthquake activity in the formation of New Zealand. The sequence of illustrations in Fig. 392 showing the development of the country during the last 50 million years is based on the removal of recently formed ocean floor and movement on fault lines. It is only one of many reconstructions.

Fig. 391. Faults active in the last 25 million years (*after Suggate, Stevens and Te Punga*).

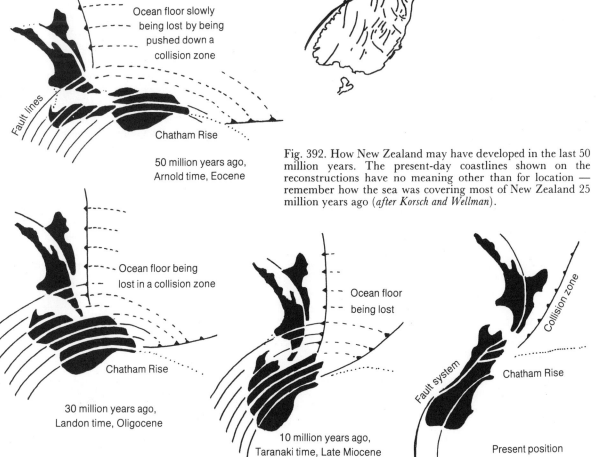

Ocean floor slowly being lost by being pushed down a collision zone

Fault lines

Chatham Rise

50 million years ago, Arnold time, Eocene

Fig. 392. How New Zealand may have developed in the last 50 million years. The present-day coastlines shown on the reconstructions have no meaning other than for location — remember how the sea was covering most of New Zealand 25 million years ago (*after Korsch and Wellman*).

Ocean floor being lost in a collision zone

Chatham Rise

30 million years ago, Landon time, Oligocene

Ocean floor being lost

Chatham Rise

10 million years ago, Taranaki time, Late Miocene

Fault system

Collision zone

Chatham Rise

Present position

197

Volcanoes, volcanic rocks and eruptions

The direction of the main fault movements illustrated in Fig. 388 matches that of the present plate boundary. This is parallel to the line of volcanic activity in the Taupo Volcanic Zone, where our active volcanoes are in fact vents for the melting crustal material formed where the Pacific Plate is moving under the neighbouring plate.

Perhaps it is now time to consider the different types of eruptions and materials to be found in our volcanic areas.

Remember that dark iron-magnesium minerals start to crystallise in the melt at very high temperatures between 1100°C and 1300°C and that pale, glassy minerals like quartz and K-feldspar do not begin to crystallise until the temperature drops to about 650°C (see page 8). The sequence applies in reverse when a rock is heated; the pale minerals begin to melt first.

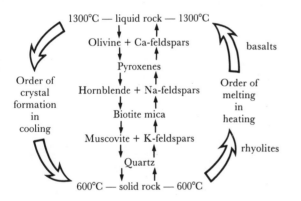

Order of crystal formation in cooling

1300°C — liquid rock — 1300°C

Olivine + Ca-feldspars

Pyroxenes

Hornblende + Na-feldspars

Biotite mica

Muscovite + K-feldspars

Quartz

600°C — solid rock — 600°C

basalts

Order of melting in heating

rhyolites

Even if the area of crust that is melting contains a fair amount of iron-magnesium minerals, if the rock reaches only 600-700°C only those minerals with a low melting point will liquify and collect together to rise to the surface as rhyolite or remain below the ground and cool slowly as granite. Because the pale, glassy minerals melt first, silica-rich magma will always be the first type of magma to form; it will become more mafic as the temperature rises and more of the iron-magnesium minerals reach their melting points.

When liquid rock collects underground in a magma chamber it is under considerable pressure and gas is dissolved in it — just as gas is

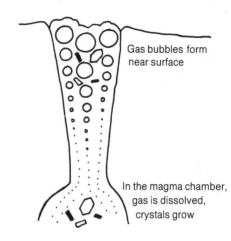

Gas bubbles form near surface

In the magma chamber, gas is dissolved, crystals grow

Fig. 393. Gas forming as magma erupts.

dissolved in beer or lemonade. As the magma churns to the surface the gas separates and forms bubbles which, like over-gassed beer, may explode into the atmosphere (Fig. 393). This is especially likely if the volcano has not erupted for some time and the vent has become plugged with solid rock. In this case the first eruption would have to be very violent in order to blow out the vent and clear a passage for later and quieter eruptions. This was seen recently at Mount St. Helens, in the United States of America, with its very violent first explosion.

In such violent eruptions the lava may completely disintegrate into thin skins of glass around bubbles of gas. The glass breaks into shards then falls as "ash" which spreads over a wide area (Fig. 394). Some fallen shards of glass retain fission tracks where radioactive particles have disintegrated; these can be used to determine the dates of past eruptions. The rock known as obsidian is the volcanic glass without the gas — the liquid part of the overflowing beer. The froth sets hard and forms an entirely different rock, pumice (Fig. 395).

Fig. 394. "Ash" is composed of glass fragments that flatten when covered with sediment.

Fig. 395. Obsidian and pumice.

The material blown out of a volcano can vary in size from small ash particles to large pumice pieces or even solid rock fragments. All the airfall material from an eruption is called tephra. Layers of tephra and the variations in size of this material may be seen in roadcuts in the Taupo area and on the Desert Road.

While vents generally direct explosions up into the air, sometimes (as at Mount St. Helens) an explosion occurs sideways (Fig. 396). In some eruptions super-heated gas and ash spew out of a volcano and over the surrounding countryside like milk boiling over, covering everything in their path with hot debris which eventually comes to rest down valleys or in depressions. There it slowly settles and welds itself into a rock which the New Zealand geologist Patrick Marshall named "ignimbrite", a firestorm. In New Zealand ignimbrite is most often seen as Hinuera Stone which is quarried for building material. It is made up of fused particles of pumice, ash, small crystals and glass fragments, many of which were flattened as the rock compacted (Fig. 397). Like lava, layers of ignimbrite crack into columnar joints as they cool, and they contain areas where the rock has not fused but remains as a loose and somewhat unstable mass.

The oldest ignimbrites are part of the Coromandel volcanics — small flows near Waihi at Owharoa and between Tauranga and Te Puke. Younger ignimbrites cover more than 25,000 square kilometres of the central North Island. Ignimbrites were first studied intensively when engineers were planning the hydro-electric dams down the Waikato River and found that all the foundations would be in ignimbrites. The cracks and unstable areas in the layers of this rock have caused some problems, first at Arapuni some 50 years ago and more recently at Wheao.

Ignimbrites are often produced by silica-rich magmas as these relatively cool masses do not release their gas easily, but tend to explode.

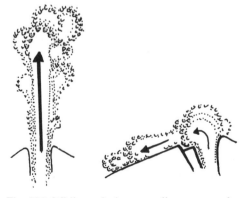

Fig. 396. While explosions usually go upwards, some may go sideways.

Fig. 397. Ignimbrite texture, in Hinuera Stone, Piarere Quarry.

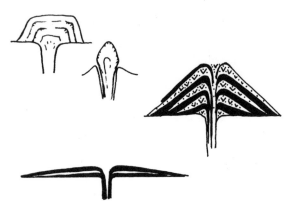

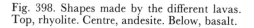

Fig. 398. Shapes made by the different lavas. Top, rhyolite. Centre, andesite. Below, basalt.

When most of the gas has been cleared a body of rhyolite lava rises sluggishly, barely hot enough to move, and oozes from the vent like toothpaste from a tube, forming a glob that may flatten on top as it cools or perhaps remain as a spine or plug (Fig. 398).

With more heat and the addition of more mafic minerals the magma mix becomes an andesite, intermediate between rhyolite and basalt. This hotter lava is more fluid than rhyolite and builds up the steep-sided cone we commonly associate with a volcano (Fig. 398). Such a cone is formed by alternating gas explosions and lava flows from a clearly visible vent or crater.

The hottest and most fluid lava of all is mafic basalt, which generally erupts quietly because gas escapes through this liquid rock. Eruptions from the basalt volcanoes of Hawaii and Iceland are tourist attractions because they are mostly non-violent events. These lavas usually fountain into the air and flow in glowing streams to the nearest depression, building up a gently sloping shield shape with a central crater (Fig. 398).

Many of our older volcanoes are so eroded that their original shape is lost or the crater or vent is filled, but volcanoes that erupted in the Pliocene and later usually have a fairly obvious shape.

The Coromandel volcanoes were still active in the Pliocene, and their last rhyolites appeared some 3 million years ago. Other rhyolite eruptions occurred in Northland, at Maungaturoto, Parahaki, Kamo and Hikurangi.

Pirongia — Karioi

A row of volcanoes stretches from Raglan inland to Tokanui (see Fig. 406 on page 206). They erupted from 3.79 to 1.8 million years ago, although the smaller cones look so fresh they could be younger.

Pirongia and Karioi are large, composite cones of thick lava flows and volcanic breccia and tuff, mainly of basalt and basaltic andesites. Pirongia, the oldest, has been very much eroded and no crater can be found; forest covers the flows on the northern sides and the southern slopes consist mainly of layers of breccia.

Mount Karioi erupted from 2.9 to 2.16 million years ago with lavas of a coarse-grained basalt to andesite, containing large crystals of feldspar, pyroxene and olivine. On its east side Karioi is bordered by smaller volcanoes with cones and flows of basalt with many olivine nodules. The latter measure up to 100 millimetres across and can be seen in the quarries at Okete and Kirikiripu, just east of Raglan. A flow from one of these smaller vents forms the ledge for the Bridal Veil Falls on the back road from Raglan to Kawhia. A 10-minute walk from the road brings you to the point where the stream plunges some 60 metres over a wall of curving, columnar basalt.

South Auckland

While the Karioi-Pirongia volcanoes were active a series of eruptions was covering the Bombay-Pukekohe area where small basalt domes, cones and explosion craters dot the landscape. The Pliocene volcanic cones are eroded, but the vents of the younger volcanoes are still visible. In this area the Pliocene basalt has weathered to produce excellent soil.

Auckland and north

In the Quaternary Period North Auckland was again peppered with eruptions of basalt, of two kinds. Horeke Basalts form the older flows that cover the landscape but have no obvious vents. They can be seen near Hokianga Harbour down the loop road to Horeke which leads to the historic Methodist Mission House at Mangungu. The younger Taheke Basalts are named from a flow by Taheke township, 18 kilometres south-west of Kaikohe. The volcanoes that produce Taheke Basalt usually have prominent cones built up of scoria — blocks of gassy, holey rock which spattered up out of the vents. They

Fig. 399. Pakaraka volcano.

all erupted probably since the last glaciation, within the last 20,000 years.

These cones made wonderful pa sites. At Pakaraka (Fig. 399), 18 kilometres north-east of Kaikohe, fortifications can be seen above fields scattered with piles of scoria boulders cleared from the land by the Maori people. European settlers later used the blocks for walls.

At nearby Ngawha Springs, hydrothermal waters associated with the last of the basalts are depositing mercury in its sulphide form — the red mineral cinnabar. Older hydrothermal deposits can be seen on the roadside near Mount Mitchell on the western loop of the Puhipuhi road about 30 kilometres north of Whangarei. The red-stained silica found here has been quarried by rockhounds, but its colour fades in sunlight so it is not a good garden rock. A little further along the road (about 100 metres) the deposit contains radiating clusters of shiny, grey, metallic needles of stibnite, antimony sulphide (Plates 8A and 8D).

In Auckland itself the volcanoes may have started erupting about 100,000 years ago, but the oldest rocks dated so far have been set at about 42,000 years, which means they would have erupted during the last glaciation.

There are more than 50 outlets for lava in the Auckland area, although very little material came up through any one vent. The youngest volcano, Rangitoto, is also the largest; the 2 cubic kilometres of basalt erupted from this source comprises more than half of all the material erupted in the Auckland field (Fig. 400).

Many of the Auckland volcanoes built up scoria cones from masses of rock spattered from their craters. In some cases the lava ran out as a flow when the gas content of the magma decreased. Some of these outlets were sited in water-soaked, low-lying rock. Here the lava and water reacted to produce enormous explosive eruptions, blowing out craters such as Lake Pupuke and Panmure Basin and leaving rings of ash or fine scoria around the vents. Occasionally, when the water had been driven off, later eruptions from the same vents produced lava flows, but these were less common.

Some of the Auckland cones have been quarried. The blocky volcanic scoria has been used for walls and gardens, and the solid basalt as a building stone (as in the Mount Eden prison) as well as a source of road metal and building aggregate. Fortunately, some of the

201

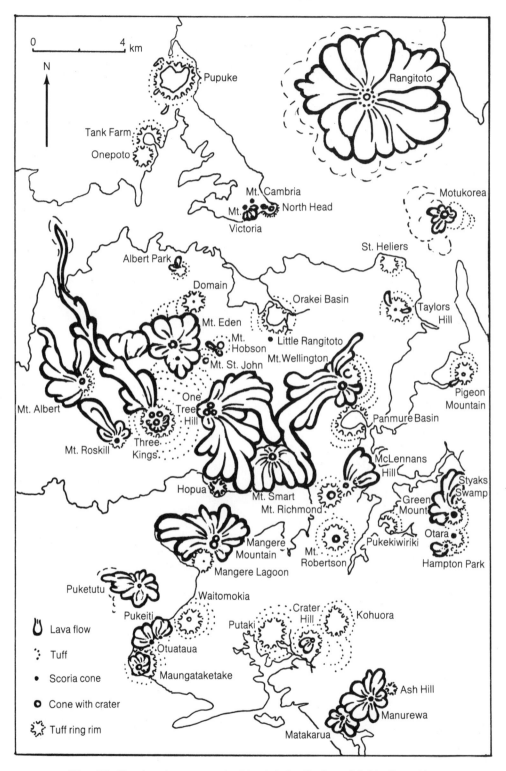

Fig. 400. Basalt volcanoes at Auckland (*after Searle, modified by Kermode*).

Fig. 401. Lava bomb, Rangitoto.

cones are reserves — for instance, Mount Eden with its crater, Maori fortifications and view of other volcanoes, including Rangitoto. The latter is a perfect example of the gentle rise of a volcano built from basalt lava flows. The little cone on top of Rangitoto is comprised of scoria.

On Rangitoto and Mount Eden it is still possible to find lava bombs, or chunks of lava that were flung into the air and cooled into distinctive spindle shapes during their fall back to earth (Fig. 401). The Auckland Museum contains larger examples.

Flows from Mount Eden and Rangitoto also produced lava tunnels. A lava tunnel is formed when a lava flow cools on the outside, leaving a tube of solid rock within which molten lava continues to flow until the lava supply is exhausted. Only the empty tube remains, ridged with wavemarks and decorated with lava stalactites. The Mount Eden tunnels are not open to the public, but those on Rangitoto may be visited.

Further information on the Auckland volcanoes may be found in E. J. Searle's book (see page 218), and in a Geological Society booklet containing background information and a detailed guide to walks on Rangitoto, Motutapu and Motuihe Islands. The Auckland Museum has a fine display on the subject.

Taranaki

If Auckland provides perfect examples of basalt volcanic shapes Mount Egmont (Fig. 402) is an example of a classic andesite cone. It is sited at the southern end of a chain of volcanoes that began to form some 1,750,000 years ago in the early Pleistocene Epoch. The Sugar Loaf Islands and Paritutu hill are dacite plugs which rose through faulted and crushed country rock; the younger volcanoes in the series are composed of andesites containing crystals of olivine, feldspar, squat augite and bladed hornblende. These can be collected in the cliffs north of Oakura Beach (Fig. 403).

Kaitake began erupting about 750,000 years ago. It has since been eroded to an area of radiating ridges, and its original andesite and diorite composition has been altered by percolating solutions to clays and quartz with small amounts of pyrite and even smaller amounts of gold, silver and copper.

Pouakai is younger, about 250,000 years old, and less worn away. Ash showers from this centre formed most of the fertile soils around Inglewood and Okato, but Pouakai was also a destructive force like Egmont, which began to build its cone about 70,000 years ago.

What we see of Egmont today may not be the first cone. All the Taranaki volcanoes are ringed with aprons of waterlaid volcanic material

Fg. 403. Oakura Beach cliffs, and detail of volcanic breccia.

which extends out to the coastal cliffs, as can be seen in the layers in the cliff across the stream at Ohawe Beach, or at Opunake Beach (Fig. 404).

Because they are built up of layers of ash and lava, andesite cones are very unstable. The loose ash and rubble erodes easily, and the layers of lava that once flowed down valleys in the ash are left as ridges. Since they are merely sitting on rubble they also collapse easily. An earthquake or gas explosion can trigger the breakdown of a wall, or an ash eruption on snow can send a mixture of melted snow, ash and rubble careering down the mountain, tearing down the valleys and picking up all the loose material in its way. Such a highly destructive mud-flow (termed a lahar) eventually stops as a pile of debris of all shapes and sizes, as mixed up as a glacial moraine; the two deposits have often been confused.

Recent mapping of the landscape round Mount Egmont has revealed a series of such flows, which have several times reached the sea. West of Egmont the plain is pimpled with small lahar mounds from both Egmont and Pouakai (Fig. 405). About 7,000 years ago the western side of Egmont collapsed to make an amphitheatre between Bobs Ridge and Fanthams Peak. Remembering the recent collapse of Mount St. Helens, when exactly the same destruction of an andesite cone occurred in the creation of vast rivers of mud and slush, we have a better idea of how Mount Egmont has been shaped and destroyed in the past and of how it has affected the surrounding countryside.

The summit plug of Egmont grew about 450 years ago after a small eruption of gas-charged ash swept down the Stony River area and burned much of the bush on the north-west slopes. One hundred and fifty years later pumice was blown into the air, and the last ash shower

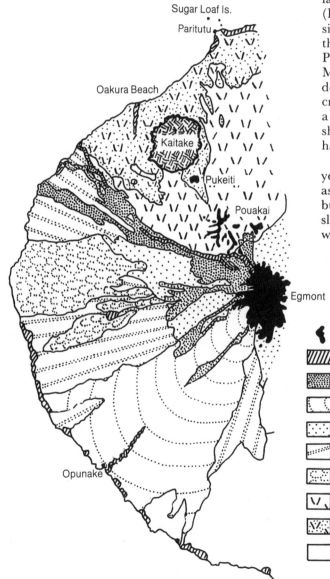

Fig. 402. Map of west Taranaki (*after Neall*).

Sugar Loaf Is.
Paritutu
Oakura Beach
Kaitake
Pukeiti
Pouakai
Egmont
Opunake

Lava	
Recent river and dune material	
Landslide and flood debris, last 500 years	
7,000-year-old lahar	
Lahars dated 500-12,000 years old	
14,000-year-old lahar	
25,000-year-old lahar with mounds	
Pouakai surfaces eroded about 20,000 years ago	
Older Pouakai surfaces	
Gravels and sandstones of southern Taranaki, about 25,000 years old	

Fig. 404. Waterlaid volcanic debris, Opunake Beach.

occurred about 230 years ago. During the same 450 years there have been at least 11 lahar flows, the most recent about 100 years ago. The National Parks Board has published a *Volcanic History of Taranaki* which describes these eruptions in more detail.

Clearly, Egmont is not yet a dead volcano and still can endanger the industries round its borders. However, the Taranaki volcanoes provided us with the raw material for industry, for the original andesite contained plentiful grains of titanomagnetite which have accumulated into the blacksands that are now mined at Waipipi, Taharoa and Waikato Heads.

Fig. 405. Lahar mounds near Opunake.

The Taupo Volcanic Zone

Although most of the earlier volcanic activity in New Zealand occurred in lines running north-west/south-east, the current major activity has changed direction and runs north-east/south-west from White Island to Ohakune, parallel to the present plate boundary to the east (Fig. 406).

The Taupo Volcanic Zone is a trough some 3.5 kilometres deep, filled with volcanic material. On each side of the trough there are faults bordering uplands of greywacke covered with flat-topped, tilted layers of ignimbrite. These ignimbrites and the ash showers associated with them account for most of the rhyolite material in the Taupo Volcanic Zone. More than 16,000 cubic kilometres of this crustal melt have poured out of the four volcanic centres in the area at Taupo, Maroa, Okataina and Rotorua.

Generally, the ignimbrites and ash showers erupted early, flowing as far as 64 kilometres from their sources and forming layers from 15 to 150 metres thick. No vents have been recognised. They are hidden because the eruption of such large amounts of material often leads to the collapse of the land above the emptied magma chamber, forming a basin called a caldera.

The formation of the Rotorua Caldera illustrates the sequence of events in an ignimbrite vent.

First, a mass of melted rock gathered in a magma chamber underground and small amounts of lava escaped to form some pumice beds (Fig. 407a). Gas pressure built up and enormous amounts of ash and pumice burst out in explosive eruptions, some in the form of

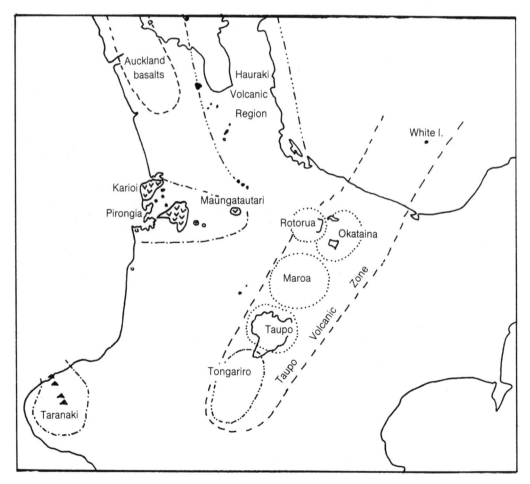

Fig. 406. Volcanic areas in the central North Island (*after Cole*).

burning clouds (nuées ardentes) which formed ignimbrites, some as towering clouds of ash which spread far across the land downwind of the vent (Fig. 407b).

As the shallow reservoir became partially emptied, the roof collapsed to form a circular basin, the caldera (Fig. 407c). The remaining lava rose up the cracks caused by the collapse on to the floor of the caldera and around the rim (Fig. 407d). A lake filled the centre part of the caldera, but its level fluctuated as other eruptions sometimes blocked the outlet. The rim towards Lake Rotoiti was covered by a much younger ignimbrite from the Okataina Centre.

The Rotorua and Okataina Centres have interacted and their flows have become interleaved. The circular caldera structure is still evident in Rotorua, but Okataina has been filled and overtopped by the rhyolite volcanoes of Haroharo and Tarawera.

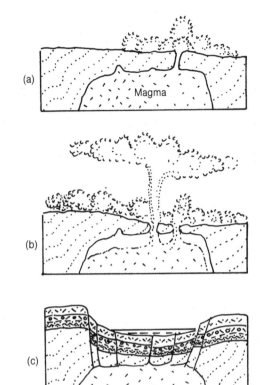

Fig. 407. Formation of the Rotorua Caldera (*after Houghton and Lloyd*).

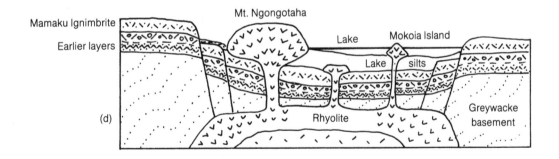

Mount Tarawera has been growing for some 20,000 years. A volcano's history may be difficult to decipher because later eruptions usually cover the older materials, but Tarawera was split open in the last eruption and the relationships of the various domes were exposed in the rift walls. The early eruptions were all of rhyolite and built up a series of flows and domes of the classic rhyolitic shape. The very first eruptions are not shown on the map (Fig. 408) but lie out to the north-east. The effects of the succeeding eruptions are shown in detail in Fig. 409.

Nine hundred years ago the Kaharoa eruptions began with an explosive clearing of the vent and ignimbrite flows. Then the Crater Dome pushed up to fill the vent (Fig. 409a). A major explosive eruption produced the Kaharoa Tephra and further ignimbrite flows which settled on top of the older flows (Fig. 409b). Finally, the Ruawahia Dome filled the tephra vent, and the Tarawera and Wahanga Domes rose in that order (Fig. 409c).

On 10 June 1886 a series of explosive eruptions opened up a chain of craters, forming an 8-kilometre rift across the mountain and blasting out blocks of basalt scoria. During the blast small lumps of rhyolite were torn from the sides of the vents in the old domes. They were coated with the black basalt of this eruption and tossed into the air, cooling as they fell to the ground. These volcanic "bombs" are unusual because of their white core and black outer skin (Fig. 410). In the crater today there are larger chunks of

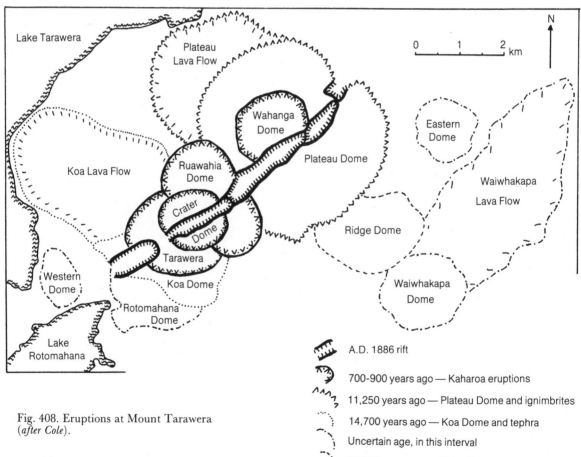

Fig. 408. Eruptions at Mount Tarawera (*after Cole*).

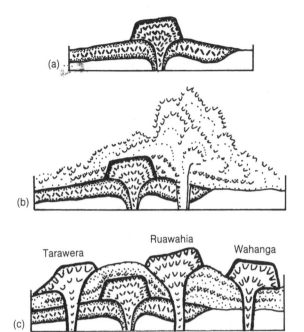

(a)

(b)

Tarawera   Ruawahia   Wahanga

(c)

Fig. 409. Events of the Kaharoa eruptions, Mount Tarawera (*after Cole*).

basalt with centres of glassy, melted rhyolite. The thick layers of red (oxidised) and black ash and scoria on top of the white material that erupted earlier make Tarawera a particularly colourful volcano (Plate 8J).

When the eruption spread south into Lake Rotomahana, violent explosions of steam spread ash and lake mud over the surrounding country-side. It was this part of the eruption that caused the widespread destruction associated with Tarawera. The devastating effects of the four-hour eruption lasted for many years.

Tarawera can be visited via the summit airstrip, by four-wheel drive vehicles or you may walk up from the car park provided. A walking track leads around the crater, and many people take a circular walk around part of the rift with an exhilarating slide down the scree slopes into the crater on the return leg. The Geological Society's guidebook *Geyserland* provides detailed coverage of Tarawera, Rotorua and nearby geothermal areas.

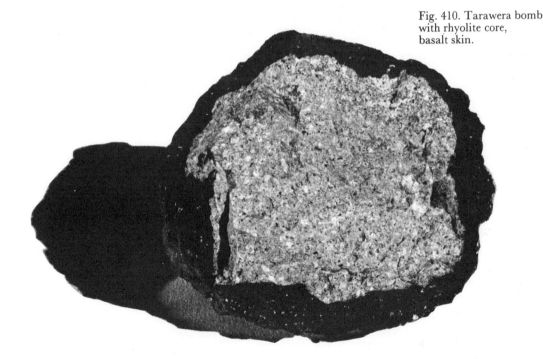

Fig. 410. Tarawera bomb with rhyolite core, basalt skin.

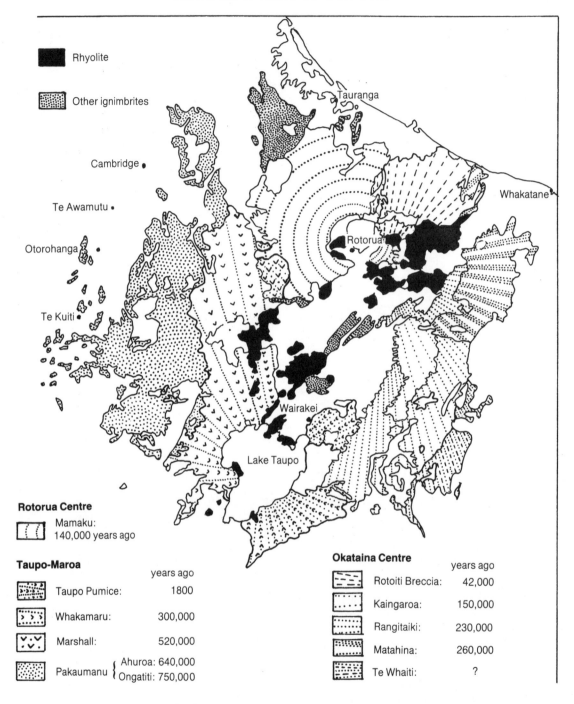

Fig. 411. Ignimbrites in the Central North Island. The dates and sources are approximate and may be altered with further research. The Pakaumanu Ignimbrite was subdivided for the area west of Lake Arapuni when detailed mapping was done and the Ahuroa Ignimbrite was found, otherwise most of it is called the Ongatiti Ignimbrite. (*After Suggate, Stevens and Te Punga*).

Fig. 412.
Ongatiti Ignimbrite
at Eight Mile
Junction, south
of Te Kuiti.

Ignimbrites cover so much of the central North Island that travellers on any of the main routes cannot help seeing them. The various layers have been mapped by their general composition and relationships, their slope and thickness (see Fig. 411) and more detailed studies are being made. The dates on Fig. 411 are very approximate — give or take 100,000 years for the older ignimbrites. Some geologists assign each layer a separate name, especially where the layers have an erosion surface or different chemistry; but in many places the detailed mapping needed to distinguish the various ignimbrites has not been done. For example, Pakaumanu Ignimbrite has been divided into Ongatiti and Ahuroa Ignimbrites west of Lake Arapuni, but not in other areas.

Ongatiti Ignimbrite was not only one of the earliest to erupt but also was among the most widespread. It flowed from Taupo or Maroa to form the cliffs above Eight Mile Junction south of Te Kuiti (Fig. 412), the low hills around Hamilton, the cliffs along Lake Karapiro and the flow that spread north towards Hinuera which is now quarried at Piarere for "Hinuera Stone" (see Fig. 397, p. 199).

At Piarere the quarry walls show two distinct layers of ignimbrite welded together, so a second flow must have arrived while the first was still warm and there must have been not one enormous explosion but several from the Taupo centre at this time. Some of these explosions would have produced towering columns of ash and pumice which were blown by the wind far across the countryside. Ash layers with fission-track ages similar to that of Ongatiti Ignimbrite are found as far apart as Oparau near Kawhia, Cape Kidnappers, Torehape in the Hauraki

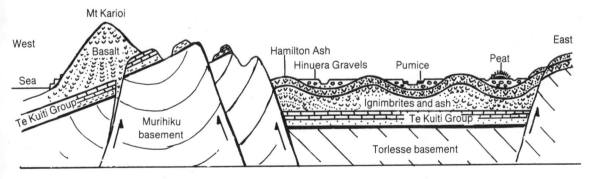

Fig. 413. Geological section through the Hamilton Basin and the Raglan hills, showing the rounded hills formed from the ignimbrites (*from Earth Science Dept., Waikato University*).

Plains and Rewa Hill near Wanganui.

In the Hamilton area towards the edge of the flow ignimbrite covered the river plain that was being eroded in the old Oligocene Te Kuiti Group sediments. The parts of the flow that were thick enough to retain their heat became welded together, but other parts eroded away, and rainfall and river action coated valley floors and lake beds in the area with almost pure pumice from this flow. The pumice was covered with river-laid debris from later eruptions, and occasional showers of ash coated the low hills in the area. Each layer of ash weathered into soils before the next layer fell. These little hills undulate across the countryside from Otorohanga north to Taupiri (see Fig. 413).

Meanwhile Whakamaru Ignimbrite erupted from Western Bay in Lake Taupo and filled in gullies in the earlier material, building up the Tokoroa Plateau. Away in the east Okataina erupted, and ignimbrites from that source flowed north and east on the Kaingaroa Plateau and the Rangitaiki River area.

The Mamaku Plateau was built up from the Rotorua Centre in a series of layers which can be seen down Leslie Road, south-east of Putaruru

(Fig. 414). In the lower part of the valley there are hummocks of the underlying older Whakamaru Ignimbrite, and above them you can see in the roadcut two distinct layers from the next series of eruptions. The third layer forms the surface of the Mamaku Plateau away to the right. Some geologists would separate the layers underlying the Mamaku Plateau into three separate ignimbrites, while others would prefer to classify them as parts of a series with a single name, Mamaku Ignimbrite. For this reason, among others, the ignimbrite map illustrated in Fig. 411 is tentative only.

During the last glacial stage the cold and windswept central plateau was covered with further ash showers and volcanic debris from the Taupo and Okataina Centres and the Tongariro-Ruapehu volcanoes. The rivers were loaded with this material. The ancestral Waikato River began to build up a plain in the basin between Maungatautari and Taupiri Gorge, which acted as a dam for the debris-choked Waikato and Waipa Rivers.

The basin was filled during two periods, the first lasting from about 65,000 years ago to about 25,000 years ago, when the Waikato River

Fig. 414. Leslie Road leaves Whites Road to Tapapa 3.5 kilometres north of Putaruru. The hummock in the valley is the older Whakamaru Ignimbrite. The roadcuts show two layers of ignimbrite, and a third flow forms the gently sloping surface of the Mamaku Plateau on the skyline.

became diverted and flowed north past the Hinuera Gap to the Hauraki Gulf (see Fig. 415). About 20,000 years ago the river resumed its old course, and for the next 3,000 years it built up a further layer of gravel, sand and silt on a braided river fan, very similar to the shingle beds made across Canterbury Plains by the Canterbury rivers.

Gravels from Waikato River built up across the little valleys in the low hills of the area, damming the older drainage channels and forming little lakes (Fig. 416). In some cases there was sufficient water movement to cut new drainage channels, but many of the lakes remained. The smaller, shallower lakes eventually became filled with peat and became

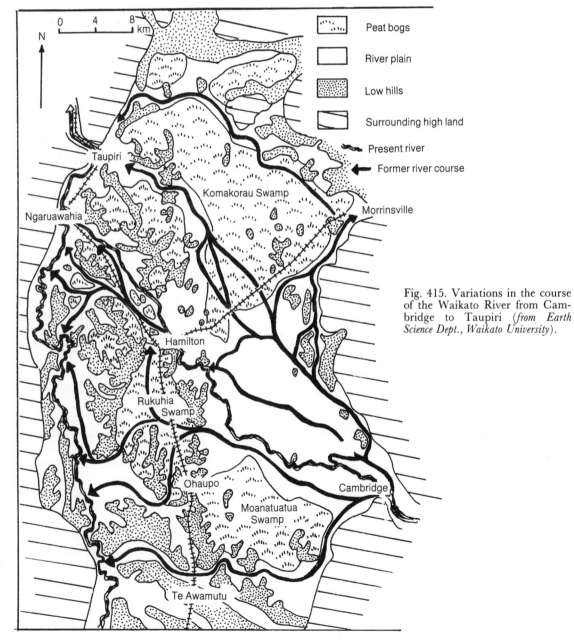

Fig. 415. Variations in the course of the Waikato River from Cambridge to Taupiri (*from Earth Science Dept., Waikato University*).

Fig. 416. View of Maungatautari across a little lake dammed by Waikato River sediments.

swamps like the large peat swamps that surround Hamilton, lying on the surface of the flood plain. The little lakes that remain can be seen along the ridge roads from Hamilton south to Te Awamutu as well as in Hamilton itself. These lakes contain sediments composed of layers of peat and ash from ash showers which record the sequence of events. This sequence can be dated from pollen and Carbon 14 in the

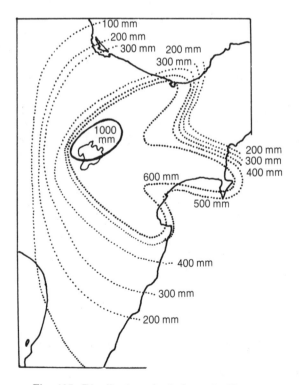

Fig. 417. Distribution of ash from the Kawakawa eruption, 20,000 years ago (*after Pullar*).

peat and from any distinct chemical characteristics in the ash. Gradually, geologists are building up a good time scale for the Late Quaternary Period in New Zealand from the sediments in the lakes of the Hamilton area.

As the amount of debris carried by the Waikato was reduced, the river began to cut down to its present bed. However, one of its terraces was constructed of pumice from the last Taupo eruption.

The Taupo volcano has been erupting ignimbrites for the last 100,000 years at widely spaced intervals. Twenty thousand years ago in one of the largest eruptions of recent time some 200 cubic kilometres of ash and pumice were thrown out, covering Napier, Gisborne and Whakatane knee-deep in ash (Fig. 417). Even in the Wellington area the ash was ankle-deep. The last eruption, of Taupo Pumice, had always been dated at about A.D.120; but recently found references in Roman and Chinese literature to the red skies and poor summers of A.D.186 are probably related to the ash in the atmosphere following such an eruption as that in Taupo. The date has accordingly been revised. In this eruption some 100 cubic kilometres of material burst out of the eastern part of Lake Taupo, perhaps from near Horomatangi Reef. (Wellington Harbour holds about 2 cubic kilometres of water, in comparison.)

The Taupo Pumice event was a series of eruptions — small explosions were followed by several large explosions which formed towering columns of ash many kilometres high. There was time for the lake to refill before the final violent explosion produced the bulk of the Taupo Pumice, followed by a brief (possibly

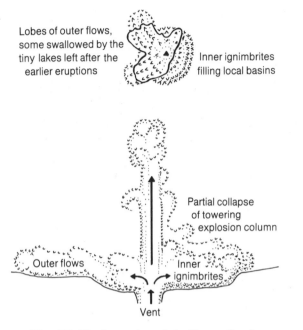

Lobes of outer flows, some swallowed by the tiny lakes left after the earlier eruptions

Inner ignimbrites filling local basins

Partial collapse of towering explosion column

Outer flows

Inner ignimbrites

Vent

Fig. 418. Final eruption of the Taupo Pumice series, A.D. 186.

two-minute) outburst of gas and ash that spread across the lake and rolled ignimbrite onto the eastern plateau (Fig. 418).

All layers of the Taupo Pumice eruptions can be seen in the walls of Earthquake Gully, where the road on the east side of Lake Taupo twists down to Rotongaio Bay. The early eruptions could have been watched from the western shores of the lake, but any observers there would have been annihilated in the last eruption when the whole area was covered with metres of the hot gas and debris, the ignimbrite.

The Taupo Pumice was washed down all the rivers arising in the area — down the Waikato onto that terrace near Hamilton, and down the Wanganui where pumice collected 4 to 6 metres deep on the river flood plain from Makirikiri to the coast.

The ignimbrites from Lake Taupo are richer in iron-magnesium minerals than those of Rotorua, reflecting their closeness to the andesite volcanoes that dominate the south end of the Taupo Volcanic Zone — Tongariro, Ngauruhoe and Ruapehu.

Tongariro is the oldest complex of the three andesite volcanoes and pebbles from this source appear in the Wanganui area in Nukumaruan

sediments, although they are more common in the Castlecliffian layers which are more than 300,000 years old. The oldest lavas so far found on the surface are about 270,000 years old; they are near the Tama Lakes between Ruapehu and Ngauruhoe (Fig. 419).

The first crater in this group may have been a large complex over the whole area, rising to a peak which has since been destroyed. The remnants of the old complex are now studded with younger craters, several of which erupted late last century.

Ngauruhoe is the latest of the craters of the Tongariro complex, built up only in the last 2,500 years. The most recent eruptions, in 1967, were of gas and ash flowing down the mountain in burning clouds (nuées ardentes); the last lava flowed down its slopes in 1954. Walking on this flow, which spread blocky lava towards the Mangatepopo hut, you can see xenoliths ("foreign stones") in the lava. They are chunks of quartz which do not belong in the lava but were torn from the basement rock.

Ngauruhoe's andesites are very dark, like basalt in composition, with visible olivine crystals as well as feldspar, augite and usually titanomagnetite. An older flow from Pukekaikiore which crosses the main road from National Park to Turangi at Mahuia Falls contains large crystalline areas of olivine (Plate 8B).

Ruapehu, the highest mountain in the North Island, rises to a summit where a main depression contains several small craters. The youngest crater contains a lake which was briefly filled with fresh lava in 1945 and has since produced a number of ash showers containing quite large chunks of lava. When a large shower lands on the snow the resulting mush (lahar) slides down the mountain valleys, destroying everything in its path. In June 1969 an eruption sent lahar flows down four major valleys, demolishing the Dome Shelter and Tea Kiosk. In April 1975 another lahar flow swept away the ski tow.

The waters from Ruapehu's crater lake escape through an ice cave to form a branch of the Whangaehu River. Several times large volumes of water have escaped down the river and caused great damage. Floods occurred in 1861, 1895 and 1925, when the railway bridge at Tangiwai was affected. Finally, on Christmas Eve, 1953, the bridge was destroyed completely

215

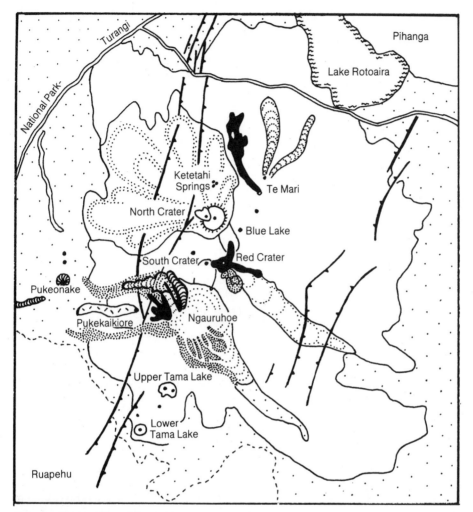

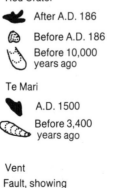

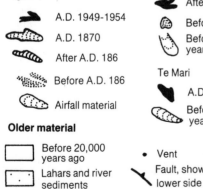

Younger flows on Tongariro

Ngauruhoe

Red Crater

After A.D. 186

Before A.D. 186

Before 10,000
years ago

A.D. 1949-1954

A.D. 1870

After A.D. 186

Before A.D. 186

Airfall material

Te Mari

A.D. 1500

Before 3,400
years ago

Older material

Before 20,000
years ago

Lahars and river
sediments

• Vent

Fault, showing
lower side

Fig. 419. Flows and vents on Tongariro
(*after Cole*).

just before the Auckland express arrived, and 151 travellers lost their lives.

Ruapehu may not be a particularly explosive volcano, but the danger it presents to the people on its slopes cannot be ignored.

Basalt lava is rarer than andesite in the Taupo Volcanic Zone, but some 40 cubic kilometres of this material have flowed out, mainly at K Trig 5 kilometres west of Taupo above Acacia Bay and at Waimarino.

By examining the chemistry of the main volcanic rocks in the Taupo Volcanic Zone geologists have tried to determine the sources of the various lavas. As the Pacific Plate descends, different areas in the collision zone are heated and produce magma. Since the chemistry of Taupo basalts is similar to that of the mantle deep below the continental crust it is thought that they may have risen from melting below the crust. The rhyolites are from melting in the crust itself, and the andesites are thought to be a mixture of the first early-melting minerals in the mantle and some of the overlying crust.

While volcanic eruptions of the central North Island are fortunately isolated outbursts, a more permanent reminder of the melted earth below may be seen in the many hydrothermal sites in the area. Here groundwaters are heated by contact with the hot earth below, boil over and discharge as hot springs or the violent plumes of geysers. The hot waters carry dissolved chemicals which crystallise in silica terraces or mounds of great beauty. Gases escaping through the wet earth turn the feldspars to clays and form mud pools.

The thermal displays at Whakarewarewa, Waiotapu, Waimangu and Orakei Korako are extensive. The first three are described in detail in the *Geyserland* guidebook (see page 219). However, there are many other smaller areas, such as those at Turangi, which reveal a little of the power in the earth below.

At Wairakei thermal power has been harnessed for electricity, and information provided at a visitors' centre explains the geology of the area. A further power station is planned for the Broadlands field to the north. The thermal region around Lake Rotokawa contains sulphur deposits which could be mined for chemicals and fertilisers, and other thermal areas are being exploited for commercial and domestic hot waters and energy; but as more water is drawn off there is a danger that the pressure will drop and thermal display areas will be spoiled.

At Wairakei and Broadlands geologists have been studying the minerals found down boreholes. The hot waters in different areas contain varying amounts of acidic or alkaline chemicals, which determine the types of metallic elements that can be dissolved in and carried by them. As the waters rise to the surface their pressures and temperatures decrease and minerals crystallise in the cracks where the waters flow. In these layers geologists have seen the same patterns of mineralisation found by earlier geologists in the ore lodes of the Coromandel. Given a few million years we could have another goldfield.

Everywhere we look today we can see how the land around us is being remodelled. Rivers are carrying their loads of sediment to swamps and to the sea, volcanoes are throwing out new igneous material, hot waters are depositing new minerals and the land is twisting and rising. These processes are forming the rocks of the future just as they made the rocks of the past.

Fig. 420. Tongariro and Ngauruhoe.

Going further

The study of rocks leads people in many directions, and this book has provided only an introduction to the subject. However, there are many books available that deal fully with the earth's minerals and the development of life on a world scale. Among these are the following less expensive and very useful titles:

Rocks and Minerals: A Little Guide in Colour, by Herbert S. Zim and Paul R. Shaffer (Hamlyn Books).

Fossils: A Little Guide in Colour, by Frank H. T. Rhodes, Herbert S. Zim and Paul R. Shaffer (Hamlyn Books).

The Hamlyn Guide to Minerals, Rocks and Fossils, by W. R. Hamilton, A. R. Woolley, and A. C. Bishop (Hamlyn Books, 1974).

The Penguin Dictionary of Geology, by D. G. A. Whitten with J. R. V. Brooks (1972).

Books on New Zealand geology include:

New Zealand Adrift, by Graeme Stevens (A. H. & A. W. Reed Ltd., 1980).

Legends in the Rocks, by Maxwell Gage (Whitcoulls Publishers, 1980).

Rugged Landscape, by Graeme Stevens (A. H. & A. W. Reed Ltd., 1974).

Bold Coasts, by Charles Cotton (A. H. & A. W. Reed Ltd., 1974).

Strata and Structure in New Zealand, by A. R. Lillie (Tohunga Press, 1980).

The Geological History of New Zealand and its Life, by Sir Charles Fleming (Auckland University Press/Oxford University Press, 1979).

City of Volcanoes, by E. J. Searle (2nd edition, Longman Paul Ltd., 1981).

The Geology of New Zealand, by R. P. Suggate, G. R. Stevens, M. T. Te Punga (Editors) (The Government Printer, 1978).

Economic Geology of New Zealand by G. J. Williams (Australasian Institute of Mining and Metallurgy, 1965 and 1974).

The volumes marked with an asterisk contain especially good lists of further reading.

Also available are the guidebooks of the Geological Society of New Zealand. These may be bought from the New Zealand Geological Survey, P.O. Box 30-368, Lower Hutt, and from some booksellers.

The most essential additions to this book are geological maps published by the New Zealand Geological Survey and available from Government Bookshops, the Science Information section of the Department of Scientific and Industrial Research or the New Zealand Geological Survey. I would begin with 1:1,000,000-scale maps of each island and add the older four-mile-to-the-inch maps for areas I planned to visit. The New Zealand Geological Survey is producing more detailed maps of some areas which are accompanied by illustrated booklets. The set began with an imperial scale of 1:63,360 and is being completed to a metric scale of 1:50,000. These new maps show rock formations and you must learn a new set of names for every map, but they are very good.

A word of caution about geological maps, including those in this book. They show the rocks geologists believe lie underneath the surface soil. They are based on the patches of rock in the area exposed by water or roading (the outcrops) and geologists' theories about the structure of the layers between the outcrops. If a rock is shown on a geological map it cannot necessarily be seen in the area, especially if it is forested or farmed. In addition, the fossils and minerals found by geologists patiently searching for means of defining the rocks may not be readily seen by the travelling public who may have only limited time to view an outcrop. In addition, the evidence that has been found is now stored away and the chances of finding further specimens depend upon the rate of erosion of new material.

This said, as far as possible the fossils and minerals illustrated in this book have been drawn from recently collected specimens or examples seen in the field. Finer and rarer specimens may be seen in university collections (especially in the museums at Victoria and Otago Universities), at the Geological Survey at Lower Hutt, in some of our regional museums and the homes of private collectors.

References

for the illustrations

Adams, C. J. D. and Oliver, P. J. "Potassium-argon dating of Mt. Somers Volcanics, South Island, New Zealand", *N.Z. Journal of Geology and Geophysics Vol. 22, No. 4* (1979): 455-63. D.S.I.R., Wellington.

Beck, A. C. Sheet 14, Marlborough Sounds (1st Ed.), "Geological Map of New Zealand 1:250,000". D.S.I.R., Wellington, 1964.

Beu, A. G.; Grant-Taylor, T. L.; Hornibrook, N. de B. Te Aute Limestone Facies. "1:250,000 N.Z. Geological Survey Miscellaneous Series Map 13". D.S.I.R., Wellington, 1980.

Bowen, F. E. Sheet 15, Buller (1st Ed.) "Geological Map of New Zealand 1:250,000". D.S.I.R., Wellington, 1964.

Bradshaw, J. D; Adams, C. J.; Andrews, P. B. "Carboniferous to Cretaceous on the Pacific margin of Gondwana: The Rangitata Phase of New Zealand" in *Gondwana Five*. Cresswell, M. M. & Vella, P. (eds). A. A. Balkema, P.O. Box 1675, Rotterdam, 1981.

Campbell, J. D; Warren, G. "Fossil localities of the Torlesse Group in the South Island", *Transactions of the Royal Society of New Zealand, Geology* 3 (8): 99-137. Wellington, 1965.

Cole, J. W. "Structure and eruptive history of the Tarawera Volcanic Complex", *N.Z. Journal of Geology and Geophysics*, Vol. 13, No. 4 (1970). D.S.I.R., Wellington.

Cole, J. W. "Andesites of the Tongariro volcanic centre, North Island, New Zealand", *Journal of Volcanology and Geothermal Research*, Vol. 3. Elsevier Science Publishers, B. V. Amsterdam, 1978.

Cole, J. W.; Lewis, K. B. "Evolution of the Taupo-Hikurangi Subduction System", *Tectonophysics*, 72: 1-22. Elsevier Science Publishers, Amsterdam, 1981.

Cooper, A. F. "Ultramafic pods from the Haast Schist Zone, South Island, New Zealand, *N.Z. Journal of Geology and Geophysics*, Vol. 19, No. 5, (1976). D.S.I.R., Wellington.

Cooper, R. A. "Lower Paleozoic rocks of New Zealand", *Journal of the Royal Society of New Zealand*, Vol. 9 (1). Wellington, 1979.

Douglas, B. J. "Preliminary paleogeographic reconstruction of fluviolacustrine setting inferred for the basal Manuherikia Group", G.S.N.Z. Annual Conference, Queenstown, Nov. 30-Dec. 4 1977 *Field Trip Guides*, Geological Society of New Zealand Miscellaneous Publications 22.

Gage, M. *The Geology of the Waitaki subdivision*, New Zealand Geological Survey Bulletin, n.s. 55. D.S.I.R., Wellington, 1957.

Gair, H. S. "The Tertiary geology of the Pareora district, South Canterbury", *N.Z. Journal of Geology and Geophysics*, Vol. 2, No. 2. D.S.I.R., Wellington, 1959.

Grindley, G. W. Sheet S8 Takaka,(1st Ed.). "Geological Map of New Zealand 1:63,360". D.S.I.R., Wellington, 1971.

Gunn, B. M. "Structural features of the Alpine Schists of the Franz Josef-Fox Glacier region", *N.Z. Journal of Geology and Geophysics*, Vol.3, No. 2, (1960) D.S.I.R., Wellington.

Harmsen, F. J. M. "The stratigraphy and sedimentology of the Awakino Gorge area, South Auckland". Unpublished B.Sc. Honours thesis, Victoria University of Wellington, 1979.

Hayward, B. W. "Eruptive history of the early to mid-Miocene Waitakere volcanic arc and paleogeography of the Waitemata Basin, northern New Zealand", *Journal of the Royal Society of New Zealand*, Vol 9: 297-320. Wellington, 1979.

Hornibrook N. de B. *Guidebook for Tour B1: Neogene Geology — North Island East Coast Basin*. 15th Pacific Science Congress and 3rd International Meeting on Pacific Neogene Stratigraphy. Dunedin, 1983.

Houghton, B. F. *Geyserland*, Guidebook No. 4, Geological Society of New Zealand. Lower Hutt, 1982.

Johnston, M. R. Geology of Nelson Urban Area, 1:25,000 (1st Ed), "N.Z. Geological Survey Urban Series Map 1". Map and notes. D.S.I.R., Wellington, 1979.

Kear, David. Sheet 4, Hamilton (1st Ed.). "Geological Map of New Zealand 1:250,000". D.S.I.R., Wellington, 1960.

Kermode, L. O. "Auckland Volcanoes", G.S.N.Z. Annual Conference 14-18 November, 1983 in *Tour Guides*. Geological Society of New Zealand Miscellaneous Publications 30, Lower Hutt.

Kingma, J. T. *Geology of the Te Aute Subdivision*, N.Z. Geological Survey Bulletin, n.s. 70. D.S.I.R., Wellington, 1971.

Korsch, R. J.; Wellman, H. W. "The geological evolution of New Zealand and the New Zealand region" in Nairn, A. E. M.; Stehli, F. G.; Uyeda, S. (eds) *The Ocean Basins and Margins*, Vol. 7A, *The Pacific Ocean*. Plenum Press, New York (in press).

Lloyd, E. F. *Geology of Whakarewarewa Hot Springs*, D.S.I.R. Information Series, no. 111, Wellington, 1975.

McBeath, D. M. "Gas-condensate fields of the Taranaki Basin", *N.Z. Journal of Geology and Geophysics*, Vol. 20, No. 1. D.S.I.R., 1977.

McKellar, I. C. Sheet 25, Dunedin (1st Ed.) "Geological Map of New Zealand 1:250,000". D.S.I.R., Wellington, 1966.

Moore, W. R. "Geology of the Raukokore area, Raukumara Peninsula". Unpublished M.Sc thesis. Victoria University of Wellington, 1957.

Mutch, A. R. Sheet 23, Oamaru (1st Ed.) "Geological Map of New Zealand 1:250,000". D.S.I.R., Wellington, 1963.

Neall, V. E. "The Volcanic history of Taranaki". Egmont National Park Board, Wellington, 1974.

Nelson, C. S. "King Country". G.S.N.Z. Annual Conference, Dec. 6-9, 1976, University of Waikato, Hamilton, *Field Excursion Guidebook*. Geological Society of New Zealand Miscellaneous Publications 21, Lower Hutt.

New Zealand Geological Survey. North Island, South Island. (1st Ed.) "Geological Map of New Zealand 1:1,000,000". D.S.I.R., Wellington, 1972.

Nicol, E. R. "Igneous Petrology of the Clarence and Awatere Valleys, Marlborough". Unpublished Ph.D. thesis, Victoria University of Wellington.

Pullar, W. A. "Isopachs of tephra, Central North Island". N.Z. Soil Bureau Maps 133/8-14 to accompany N.Z. Soil Survey Report 1. D.S.I.R., Wellington, 1972.

Searle, E. J. "Auckland volcanic district", in *New Zealand Volcanology. Northland, Coromandel, Auckland, N.Z.* D.S.I.R. Information Series 49, Wellington, 1965.

Stevens, G. R. *Rugged Landscape*. A. H. & A. W. Reed Ltd., Wellington, 1974.

Stevens, G. R. *New Zealand Adrift*. A. H. & A. W. Reed Ltd., Wellington, 1980.

Stevens, G. R. *The anatomy of a Marlborough fault line; the Wairau Fault at Branch River*. Guidebook No. 2, Geological Society of New Zealand, Lower Hutt, 1975.

Suggate, R. P.; Stevens, G. R.; Te Punga, M. T. (eds) *The Geology of New Zealand*. Government Printer, Wellington, 1978.

Watters, W. A.; Speden, I. G.; Wood, B. L. Sheet 26, Stewart Island (1st Ed.), "Geological Map of New Zealand 1:250,000". D.S.I.R., Wellington, 1968.

Weaver, S.; Sewell, R.; Dorsey, C. *Extinct Volcanoes: A Guide to the Geology of Banks Peninsula*. Guidebook No. 7. Geological Society of New Zealand, Lower Hutt, 1985.

Wellman, H. W. "Divisions of the New Zealand Cretaceous". *Transactions of the Royal Society of New Zealand*, Vol. 87 (1-2):99-163. Wellington, 1959.

Wilson, D. D. *Geology of the Waipara Subdivision*. N.Z. Geological Survey Bulletin, n.s. 64. D.S.I.R., Wellington, 1963.

Wood, B. L., *Geology of the Tuatapere subdivision, western Southland*. N.Z. Geological Survey Bulletin, n.s. 79. D.S.I.R., Wellington, 1969.

Wood, B. L. Sheet 22, Wakatipu (1st Ed.) "Geological Map of New Zealand 1:250,000". D.S.I.R., Wellington, 1962.

The cross-sections of the Banks Peninsula volcanoes were drawn from student field guides supplied by the Geology Department, University of Canterbury. The map and cross-section of the Hamilton area were from student field notes supplied by the Department of Earth Science, Waikato University.

Index

Colour plate references are indicated in bold type. Entries for individual fossils, minerals and rocks will be found under these category headings.

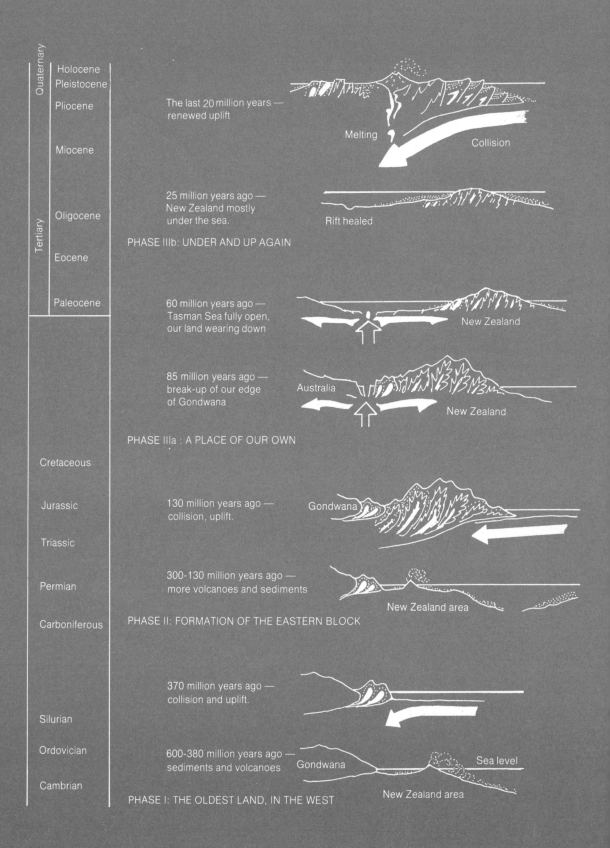

Quaternary	Holocene	
	Pleistocene	
	Pliocene	The last 20 million years — renewed uplift
	Miocene	
Tertiary	Oligocene	25 million years ago — New Zealand mostly under the sea.
		PHASE IIIb: UNDER AND UP AGAIN
	Eocene	
	Paleocene	60 million years ago — Tasman Sea fully open, our land wearing down
		85 million years ago — break-up of our edge of Gondwana
		PHASE IIIa : A PLACE OF OUR OWN

Melting · Collision

Rift healed

New Zealand

Australia · New Zealand

Cretaceous

Jurassic — 130 million years ago — collision, uplift.

Gondwana

Triassic

Permian — 300-130 million years ago — more volcanoes and sediments

New Zealand area

Carboniferous — PHASE II: FORMATION OF THE EASTERN BLOCK

370 million years ago — collision and uplift.

Silurian

Ordovician — 600-380 million years ago — sediments and volcanoes

Gondwana · Sea level

Cambrian

PHASE I: THE OLDEST LAND, IN THE WEST

New Zealand area